DANGEROUS
ANIMALS

DANGEROUS ANIMALS

BARRY MADDEN

amber
BOOKS

This pocket edition first published in 2026

First published in 2022

Published by
Amber Books Ltd
United House
London N7 9DP
United Kingdom
www.amberbooks.co.uk
Facebook: amberbooks
YouTube: amberbooksltd
Instagram: amberbooksltd
X(Twitter): @amberbooks

ISBN: 978-1-83886-677-8

Editorial, design and picture research: Amber Books Ltd

Printed in China

Contents

Introduction

We share this world with a multitude of wonderful creatures, from tiny insects that go about their lives largely unnoticed, to huge marine mammals that enthrall us with their sheer size and majesty. Between the covers of this book, you will be introduced to some of the world's most impressive and dangerous animals. Impressive for the skills they possess as superbly evolved hunters, and masters of their environment; dangerous in the sense that they can be ruthlessly efficient predators, possessing speed, strength and acutely tuned senses. Unfortunately, most of the animals you will read about here are under threat from habitat destruction, poaching, human encroachment and ignorance. Collectively we need to reconnect, recognizing not just the great importance of nature for its own sake, but also for our own wellbeing. We hope this book inspires you to look afresh at the wild, and will nurture your desire to help conserve wild places and wild things so that the creatures featured herein, together with thousands of others around the globe, can prosper.

OPPOSITE:

Great white shark

Great white sharks are notable for their size, with larger females growing up to 6.1 metres (20ft) in length.

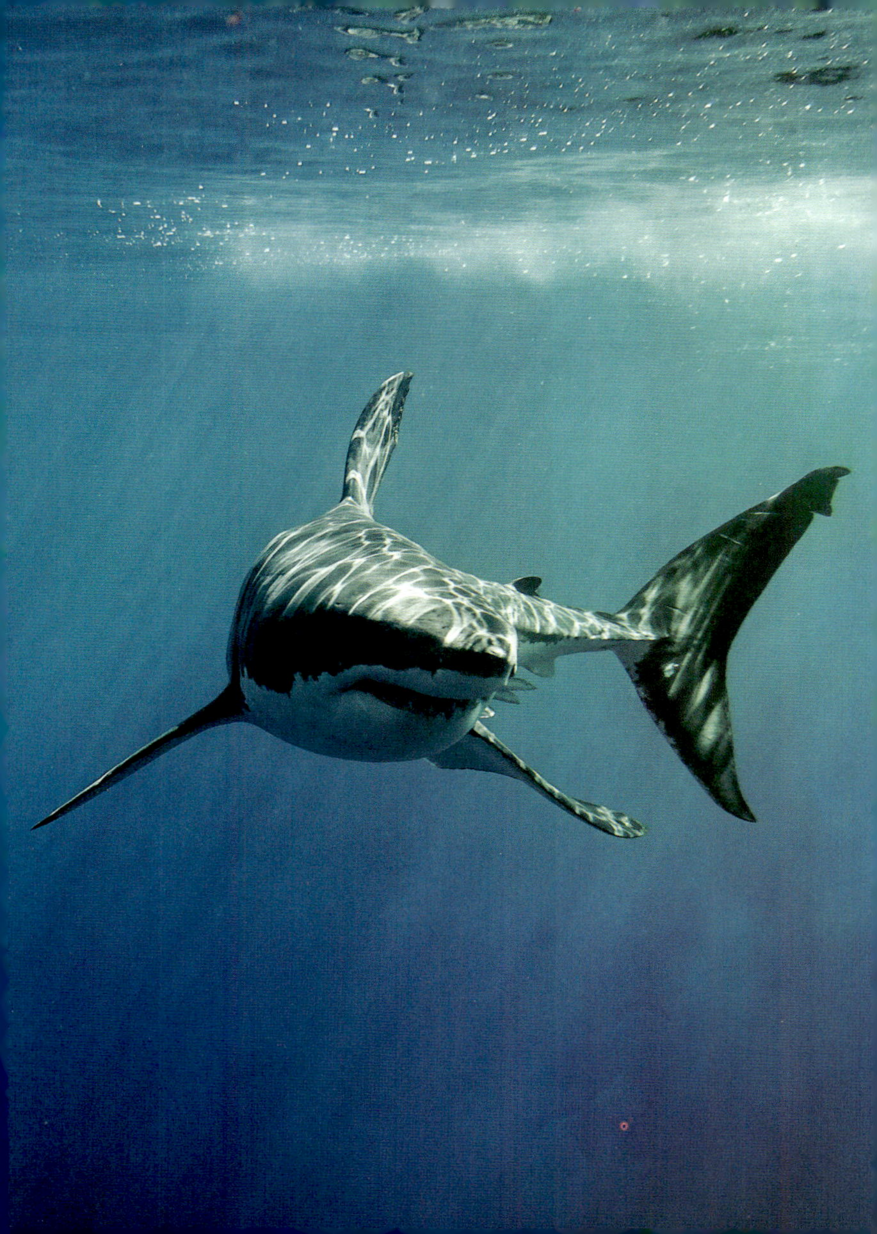

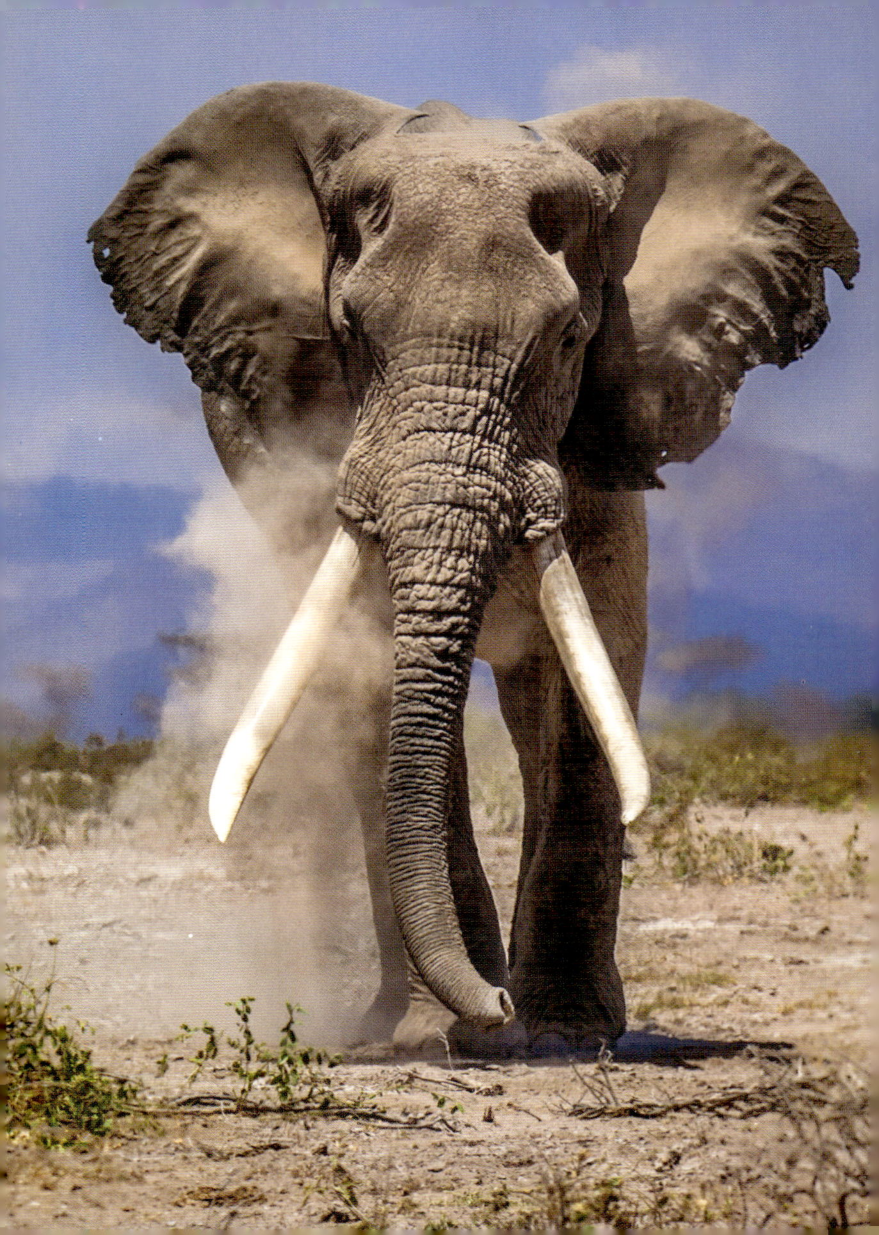

Africa

Africa! Alive, pulsating, vital. The smells and the sounds, the colours and sense of space. Spectacular scenery, sensational wildlife encounters; it is simply a wonderful place. Put simply, Africa gets under your skin. There's nothing on earth quite like it. Maybe it's the simple happiness of the people, always smiling, always radiant. Maybe it's the humbling effect of being surrounded by so much animal life. Maybe it's the realisation that we are, individually, a small, insignificant being lost in the immensity of nature. Whatever, a visit will humble, amaze, delight and surprise. The continent of Africa is a massive landmass containing a wide range of climatic zones and associated wildlife. From the parched deserts of the Sahara, equatorial rainforests of the Congo Basin and seasonally luxuriant savannah grassland to the temperate zones of the south, it reveals a multitude of interesting creatures that have adapted to living in extremes of temperature, drought, famine and abundance. Some, such as the African elephant, are familiar and revered; others, particularly top predators like lions, are respected and feared. But there is so much more to see on this continent of amazing diversity. Sadly, many of the species featured here are threatened, mainly by humankind, but with enlightenment there is hope for a brighter future.

OPPOSITE:

Bull African elephant

A giant of the African bush, there are two recognised species of African elephant, the African bush (savannah) elephant (as shown here), and the smaller African forest elephant. They use their trunks to suck up water and to gather food, and their strong tusks to strip bark, dig for water or fight.

Herd of female and young African elephants
These intelligent, sensitive, highly social animals roam widely across Africa. Adult females and young animals travel in herds led by a matriarch. The picture shows clearly how young animals are protected by the herd. Adult bulls generally travel alone or in small groups. African elephants can live for up to 70 years.

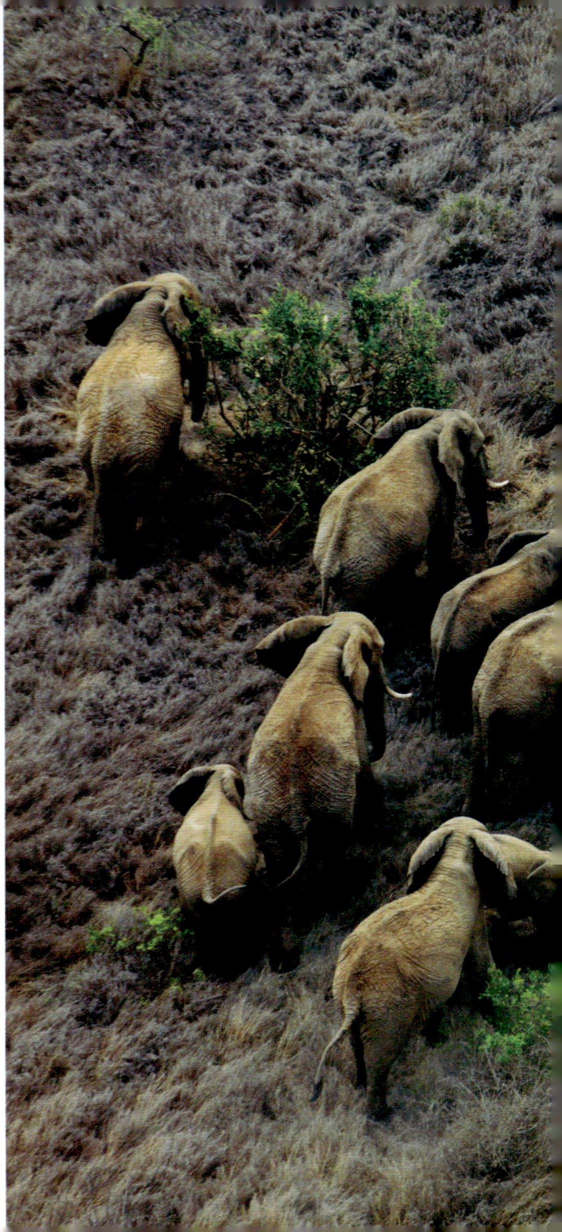

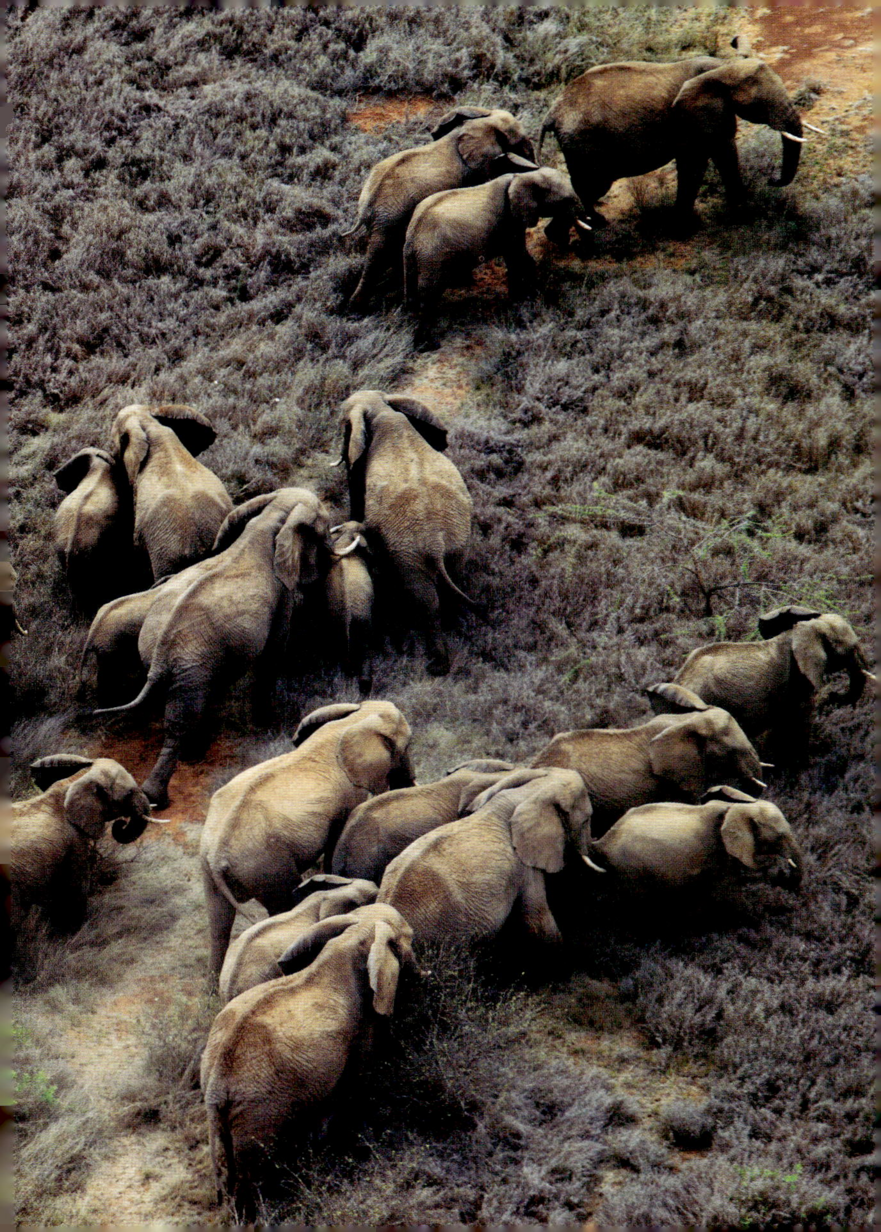

Bull elephants fighting
Weighing in excess of 5.5 tonnes (5.4 tons) and reaching a shoulder height of up to 4m (13ft), bull elephants are impressively strong animals. Here, two mature bulls tussle for mating rights with a group of females. Fighting is intense and sometimes results in serious injury or death.

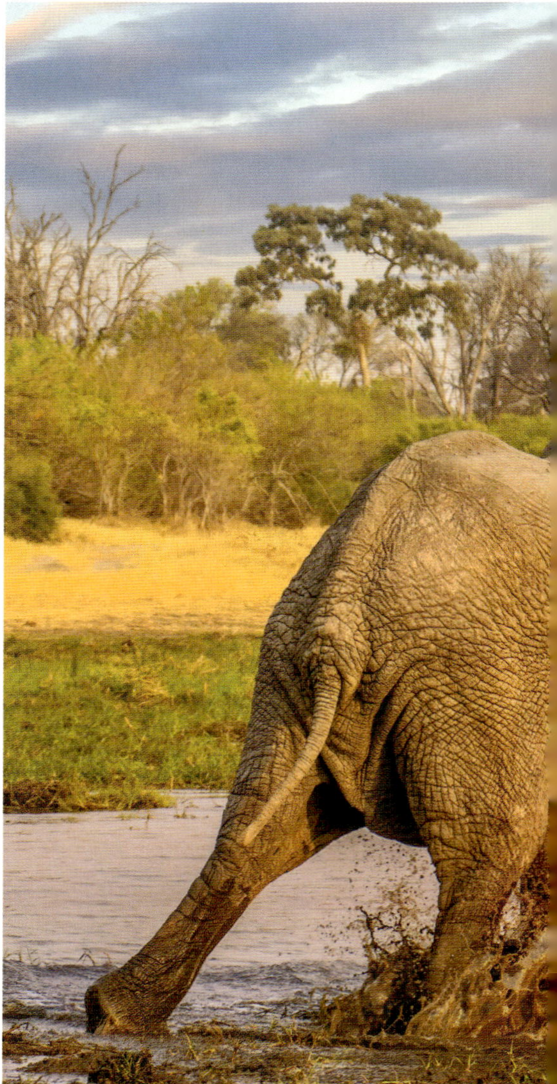

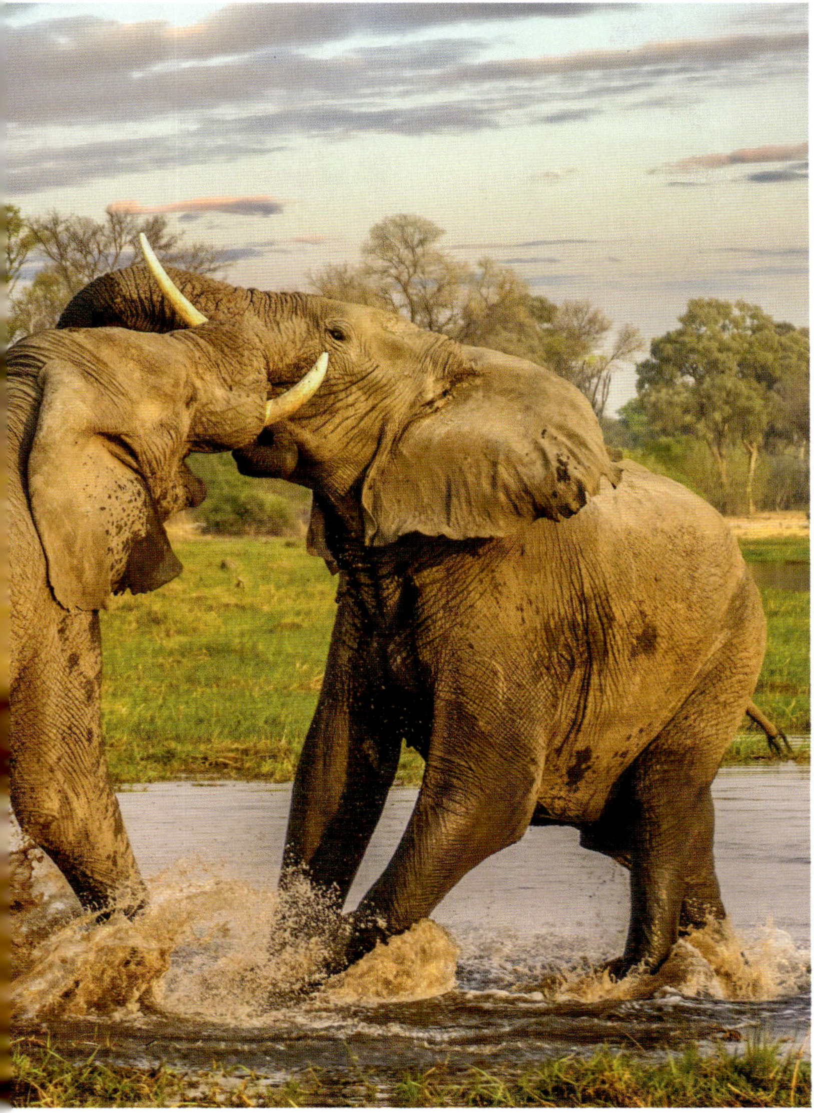

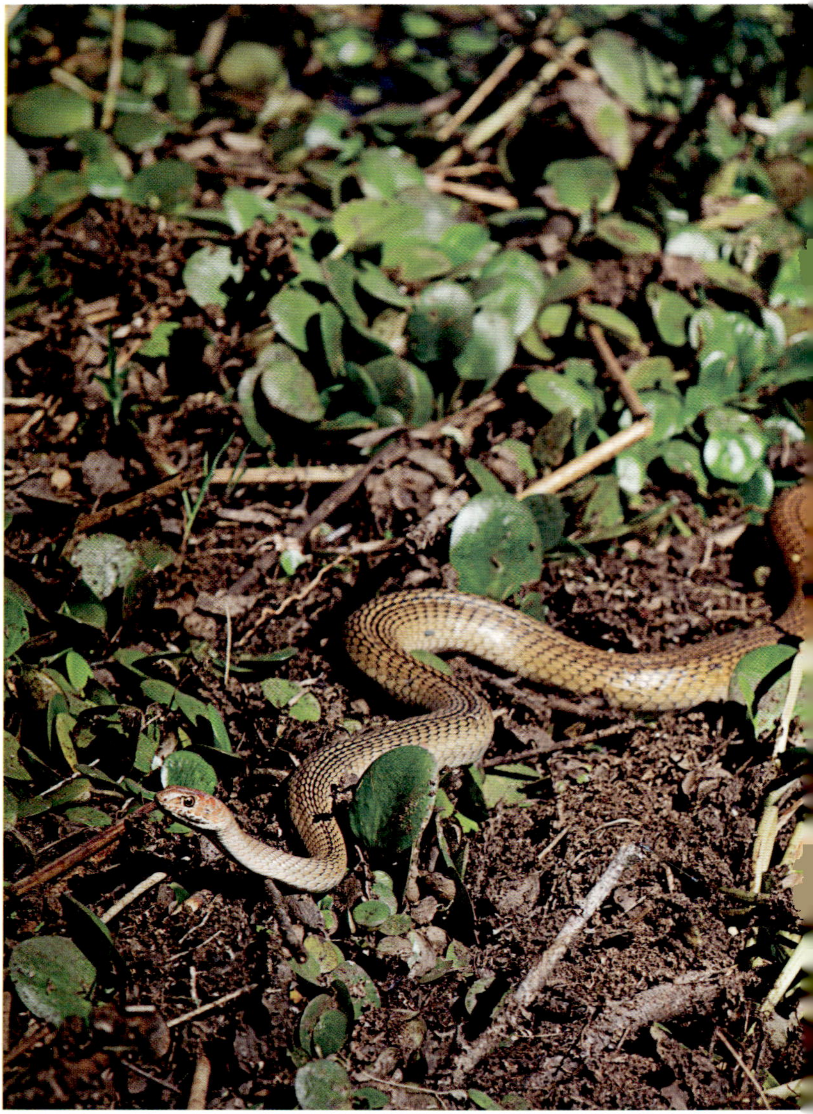

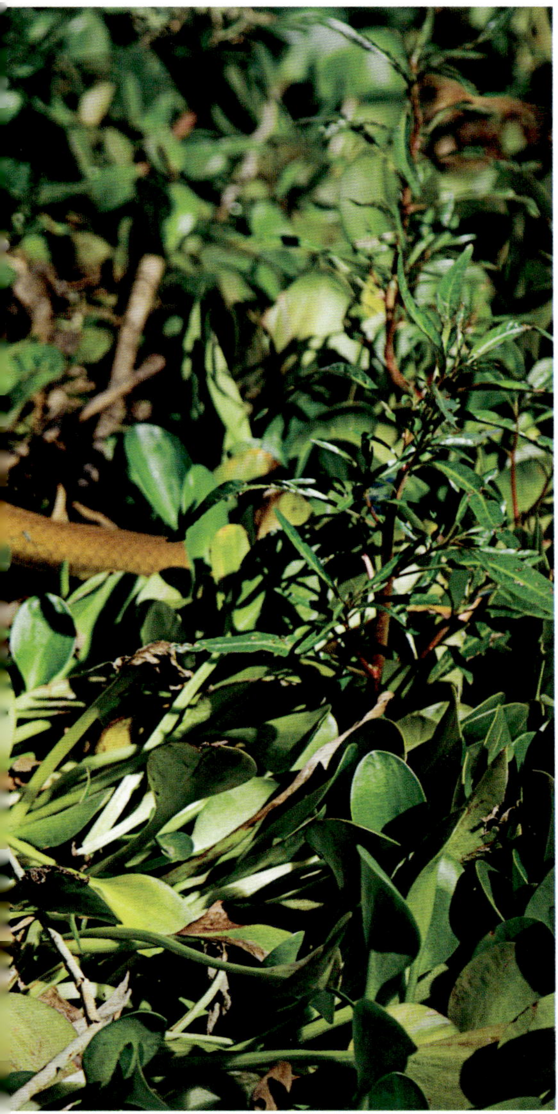

Black mamba

These snakes inhabit large areas of sub-Saharan Africa. They are Africa's longest venomous snake, reaching lengths of over 3m (10ft). Black mambas are highly venomous and can move fast. Their main prey is small mammals and birds, which they hunt on the ground or in trees.

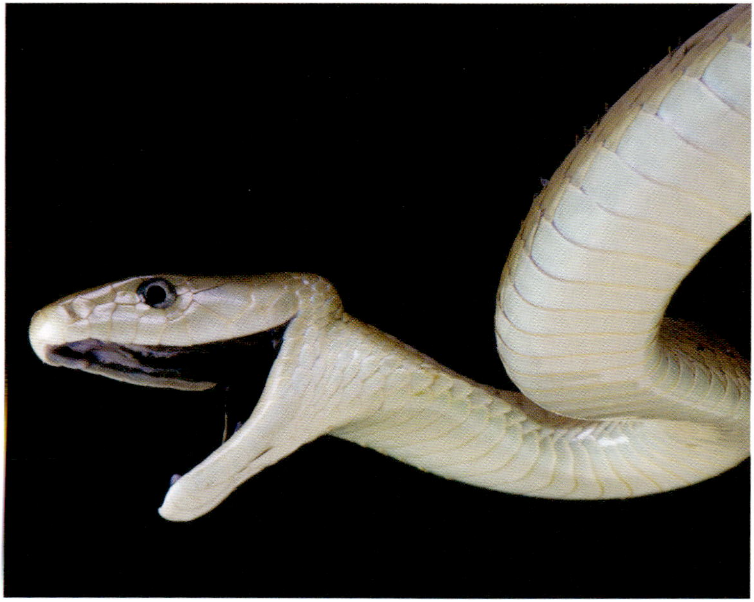

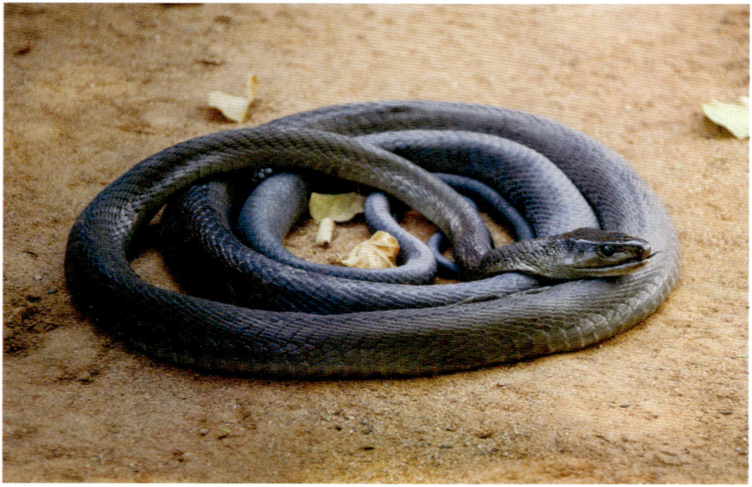

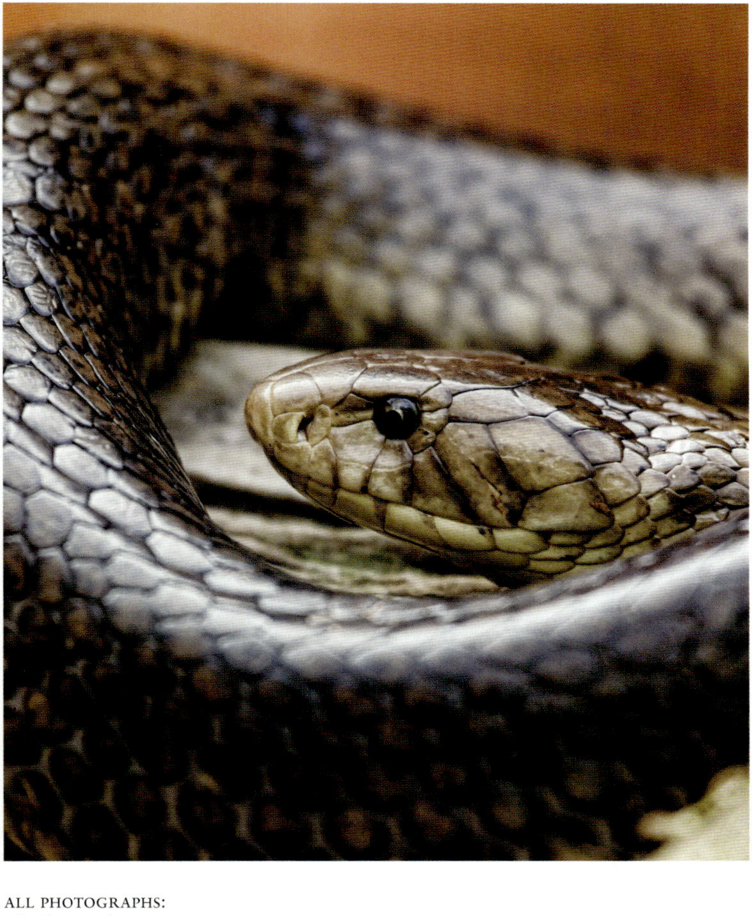

ALL PHOTOGRAPHS:

Black mamba

So-called from the black coloration inside the mouth shown during threat display, these snakes are actually greenish-brown in colour. Despite their fearsome reputation, the black mamba prefers to avoid contact with humans and will quietly slip away well before any confrontation.

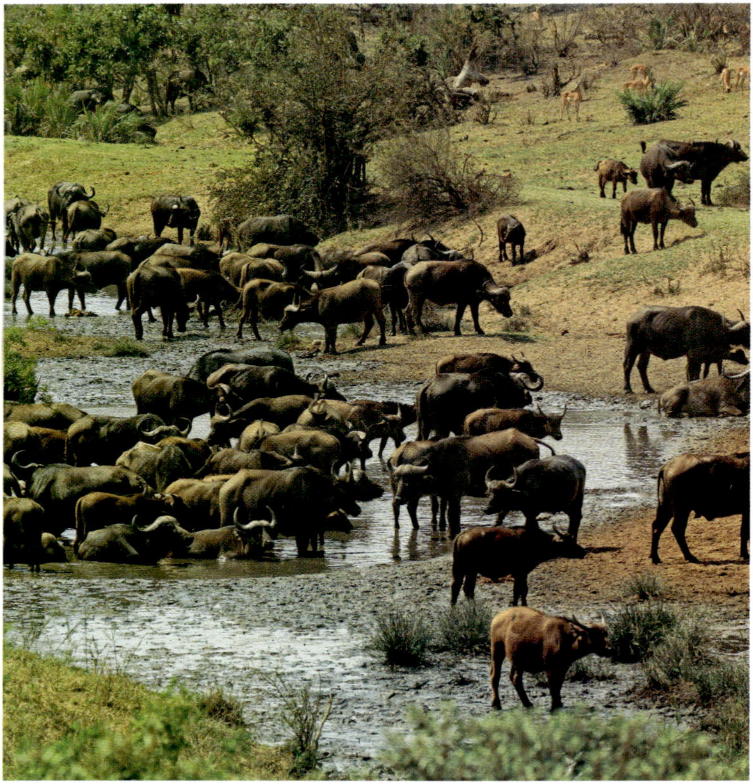

ALL PHOTOGRAPHS:

Cape buffalo

Cape buffalo are large animals that reach lengths of up to 3m (10ft) and weigh in excess of 850kg (1,874lb). They form large herds across wetlands and grasslands of sub-Saharan Africa. Prized as one of the 'big five' species, they are a sought-after sighting by safari groups. Buffalo can be dangerous and unpredictable, for this reason they have never been successfully domesticated. Young bulls often engage in mock fights, sparring with an opponent in a show of strength. Fights between mature bulls are to establish dominance within the herd, when the solid mass of horn on their heads, the 'boss', is used to push an opponent off balance.

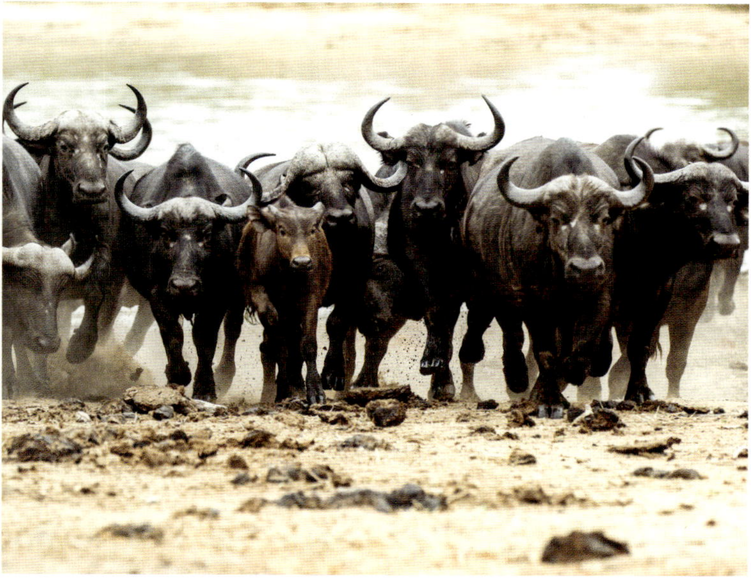

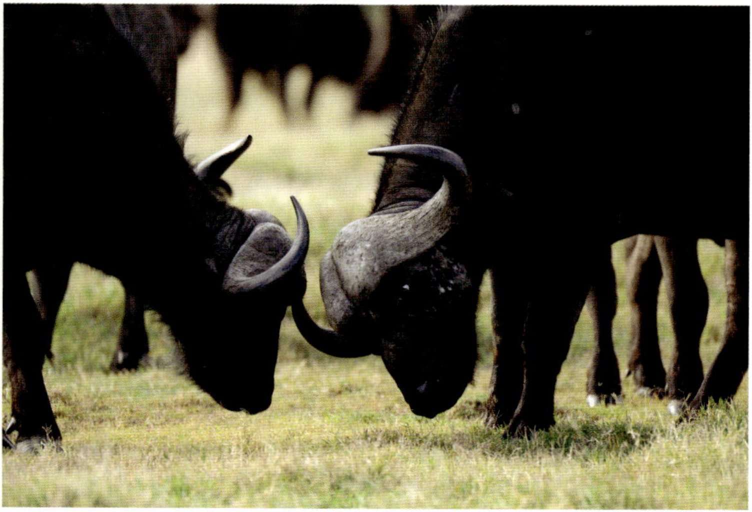

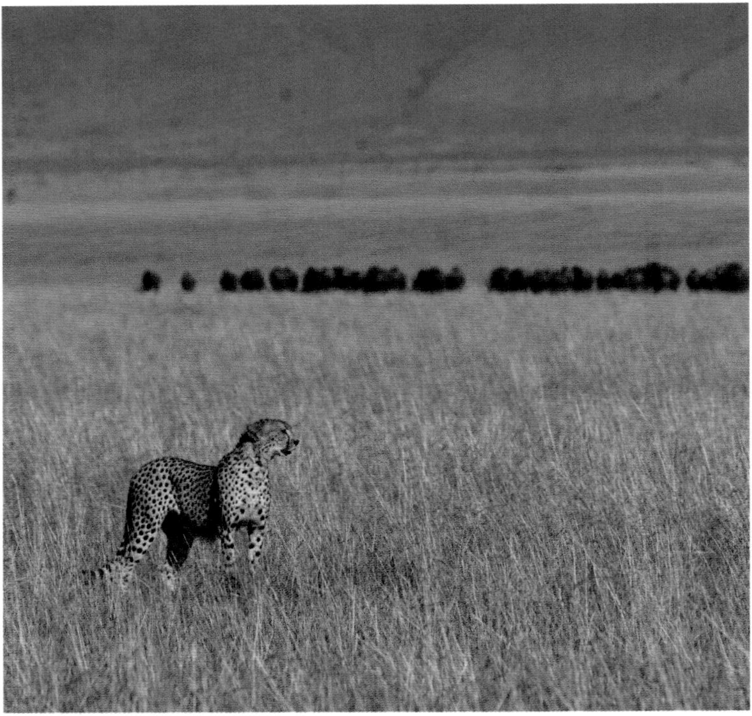

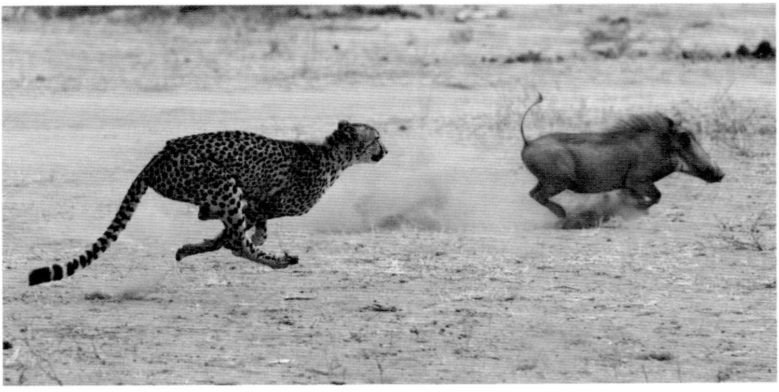

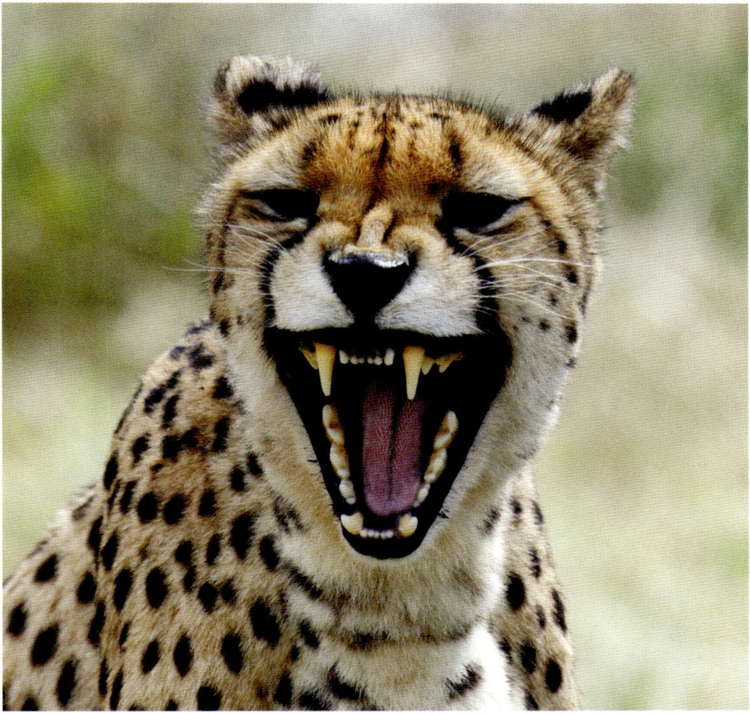

OPPOSITE TOP:
Cheetah surveying its domain
Built for speed, this sleek, lithe animal is one of Africa's top predators. It inhabits open savannah often using raised vantage points from which to locate and select prey. Cheetahs prefer a solitary existence, although males will sometimes form small collaborative groups.

OPPOSITE BOTTOM:
Cheetah chasing a warthog
Cheetahs hunt small- to medium-sized animals such as gazelle and warthogs. Here, a cheetah has surprised a warthog and is using its powers of acceleration to catch its prey. Cheetahs can reach speeds of 93km/h (58mph), although this is only sustained for short periods.

ABOVE:
Cheetah
Some notable characteristics of a cheetah can be seen here: small round head, long canine teeth, short, rounded ears and a short snout with distinctive black facial streaks running down from the corner of the eye to the mouth.

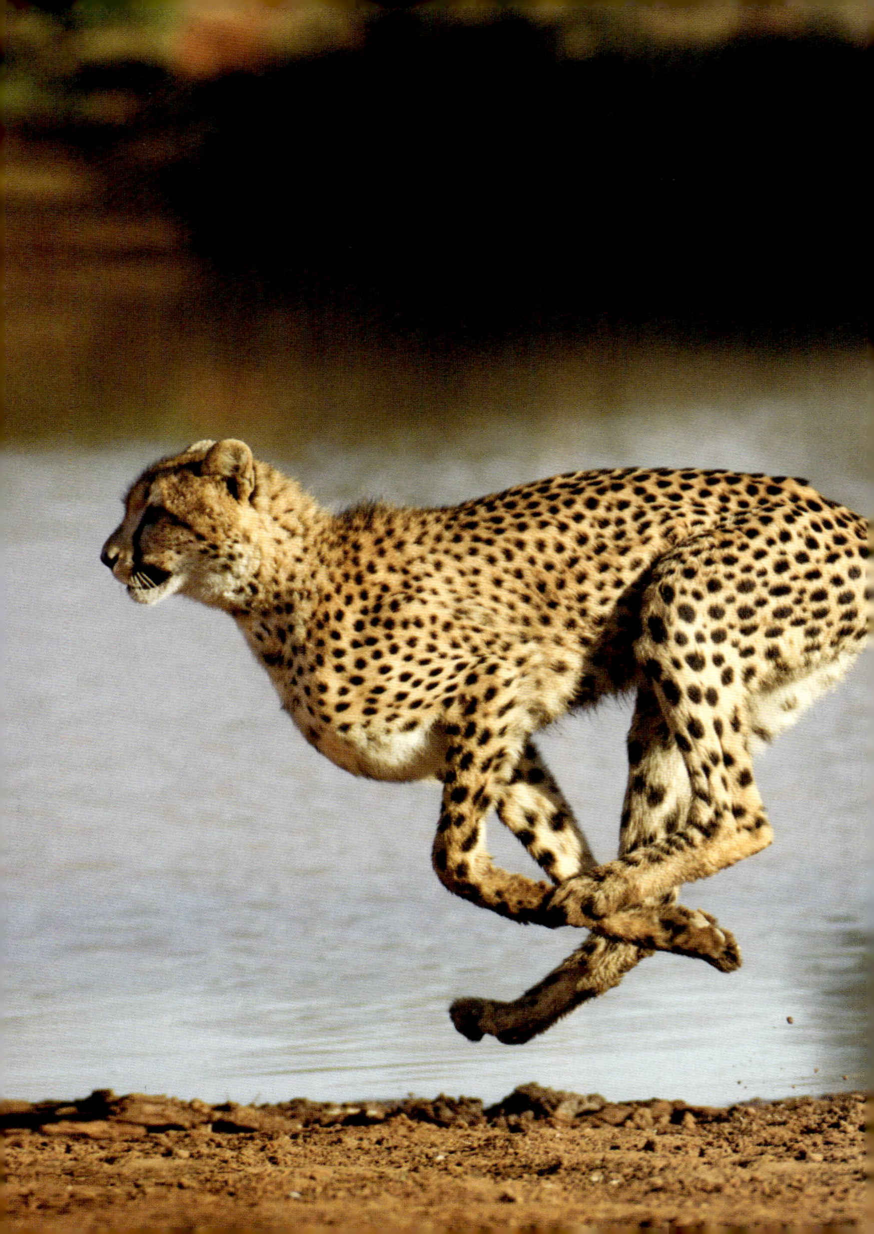

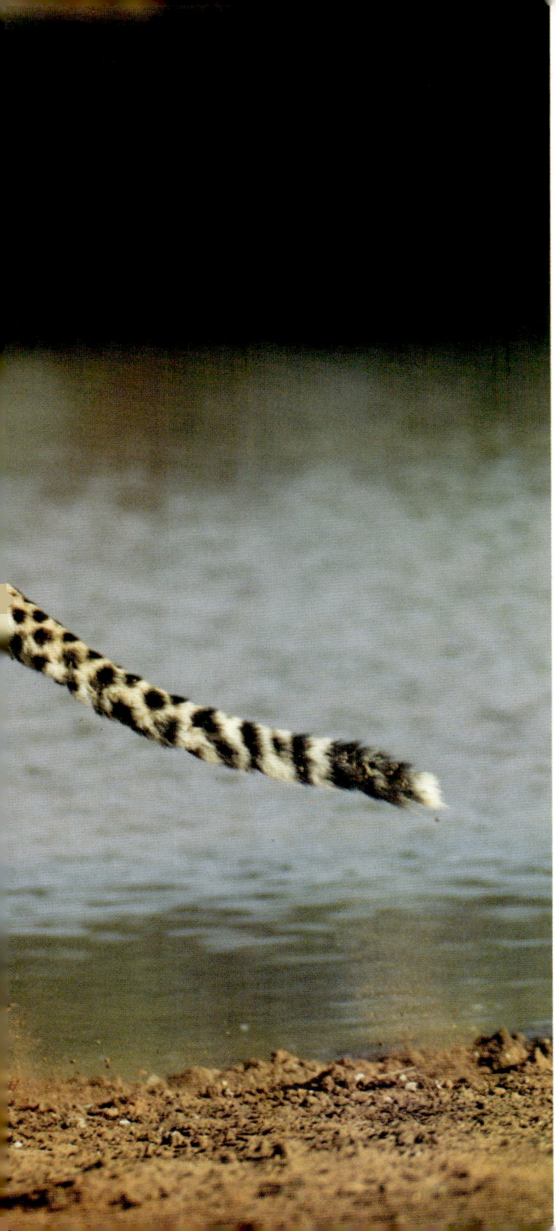

Cheetah running at high speed

Cheetahs have long legs and heavily weighted tails to provide maximum manoeuvrability when hunting at high speed. The body is covered in small round, black spots forming a pattern unique to the individual. Cheetahs typically reach shoulder heights of 70–90cm (27–35in), with a body length of 115–150cm (45in–59in).

RIGHT:

Fat-tailed scorpion

A resident of the desert regions of North Africa, this scorpion, measuring 7–9cm (2.7–3.5in), is one of the deadliest on earth. In fact, its scientific name 'Androctonus' roughly translates from Greek as 'man-killer'. If untreated, its venom will attack the central nervous system and can cause death within a few hours.

OPPOSITE:

Chimpanzee

Chimpanzees are our closest living relatives. They occupy tropical areas of Africa where they live in hierarchical social groups governed by a dominant male. Groups are highly territorial and will aggressively defend their patch from other troops. Chimpanzees communicate by using facial expressions and various vocalisations.

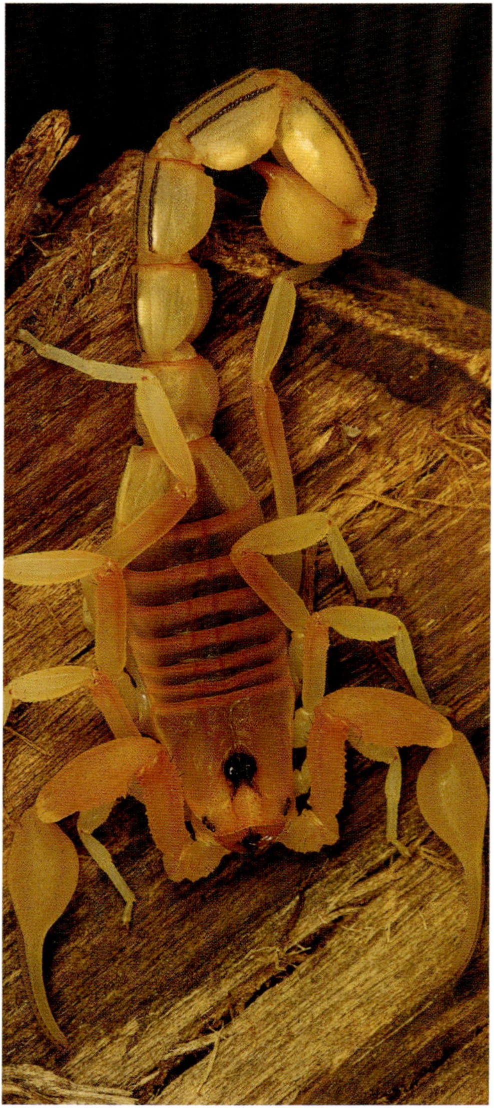

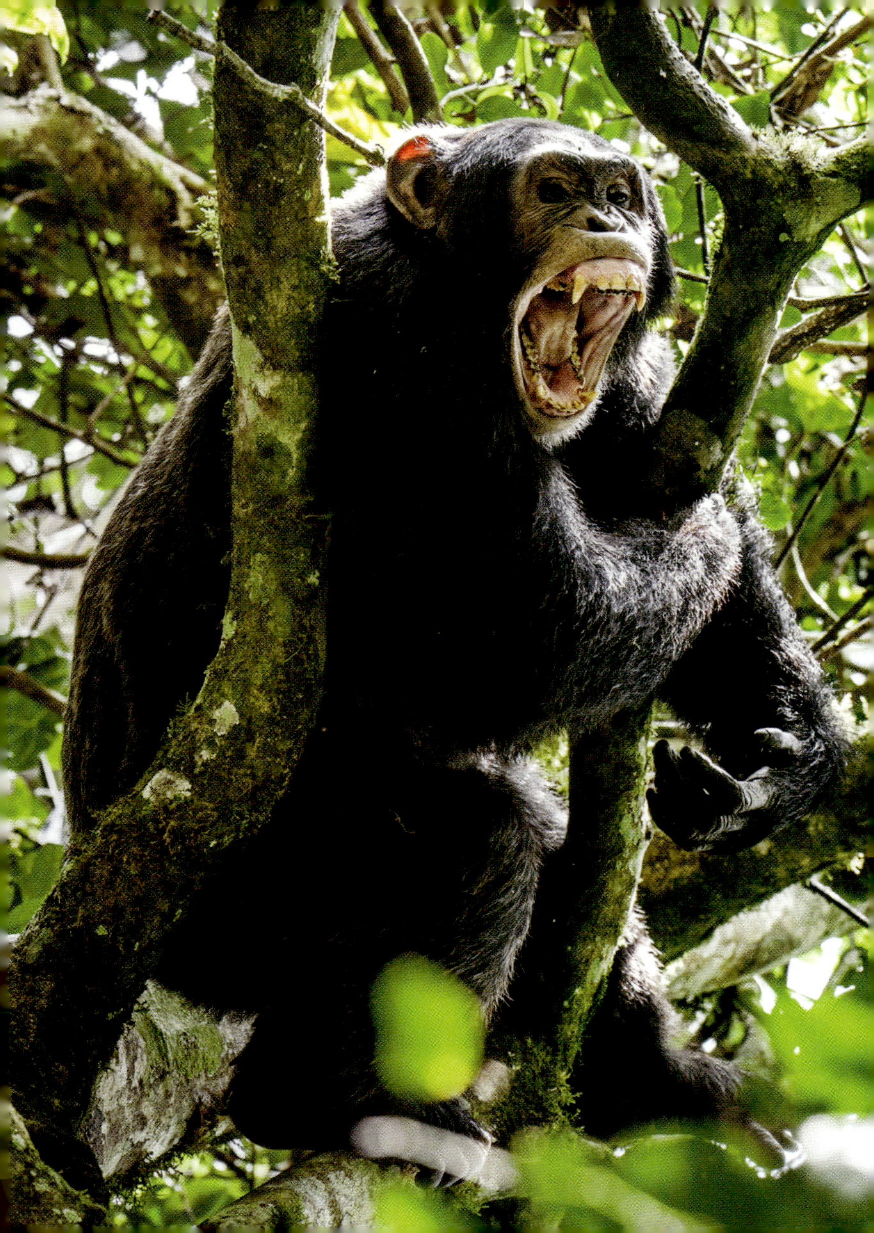

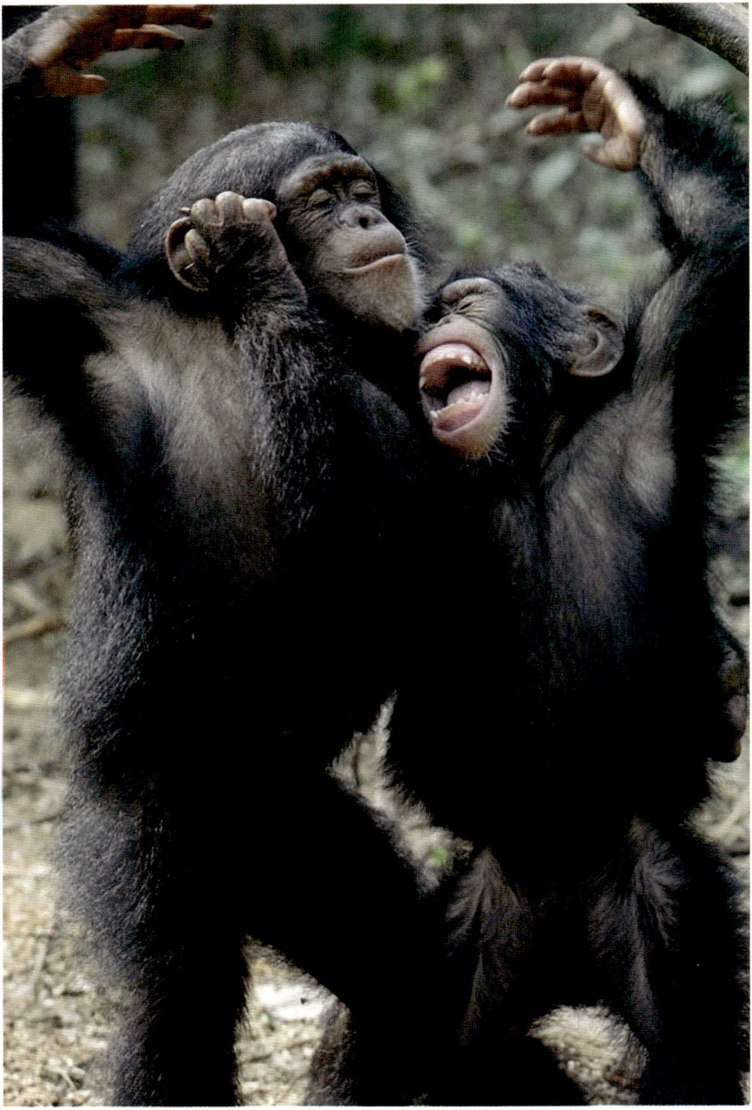

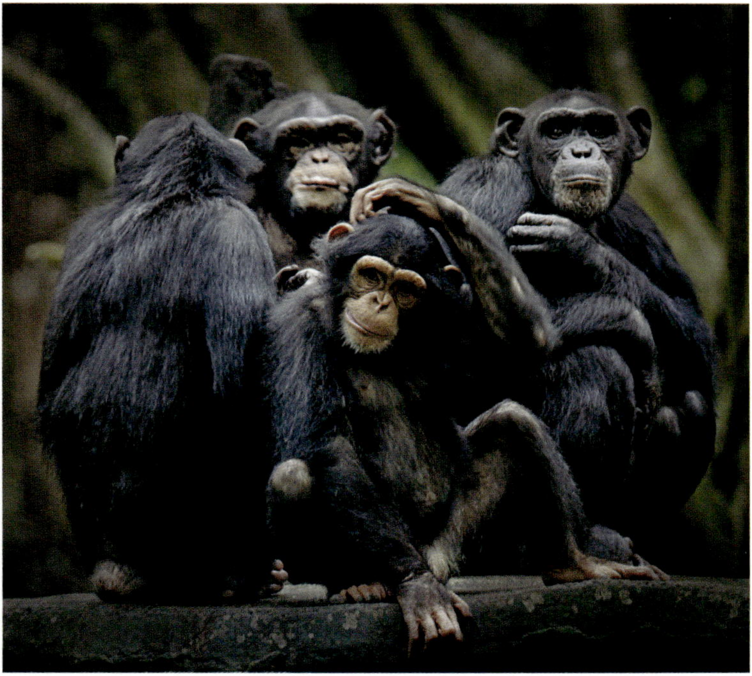

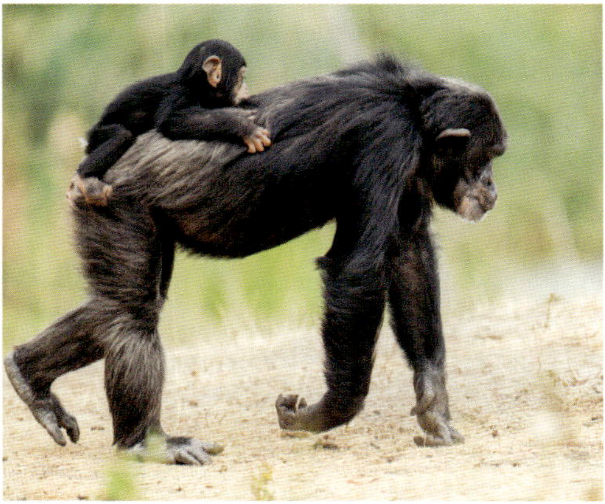

ALL PHOTOGRAPHS:

Chimpanzee

Male chimpanzees weigh 40–70kg (88–154lb) and females 27–50kg (60–110lb). Standing height is 120–150 cm (4–5ft). They can climb well, but can also 'knuckle walk'. They are primarily fruit eaters, with their large canine teeth assisting with their wider diet of bark, leaves, birds and smaller mammals.

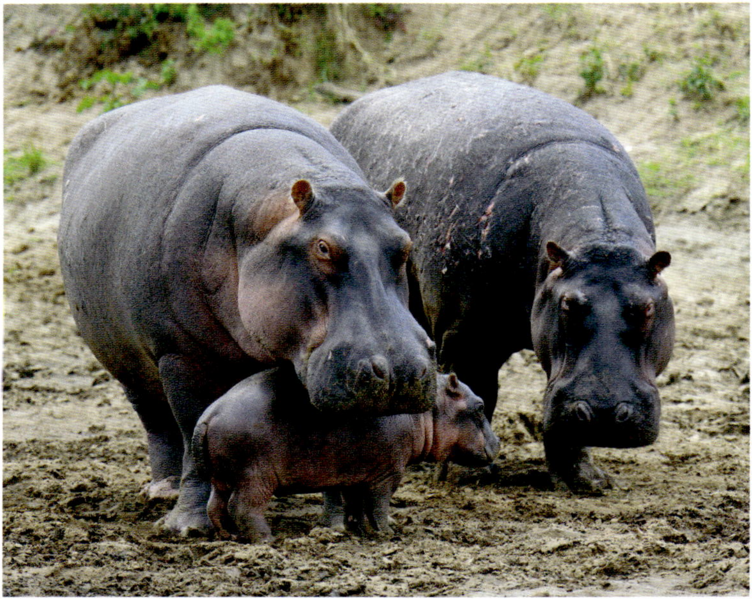

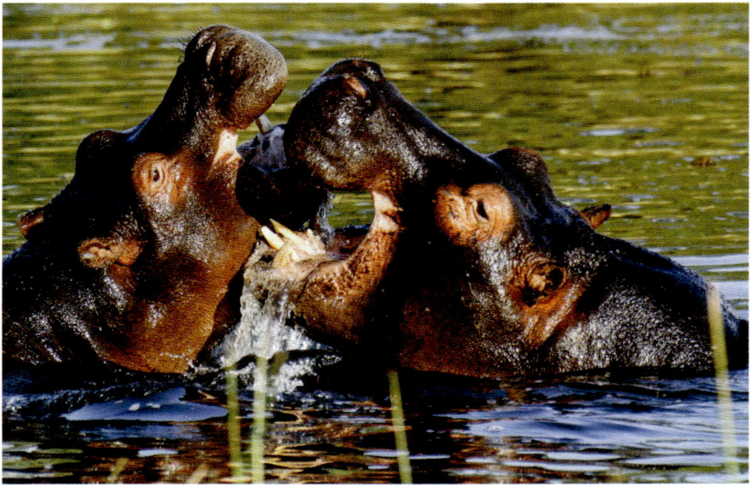

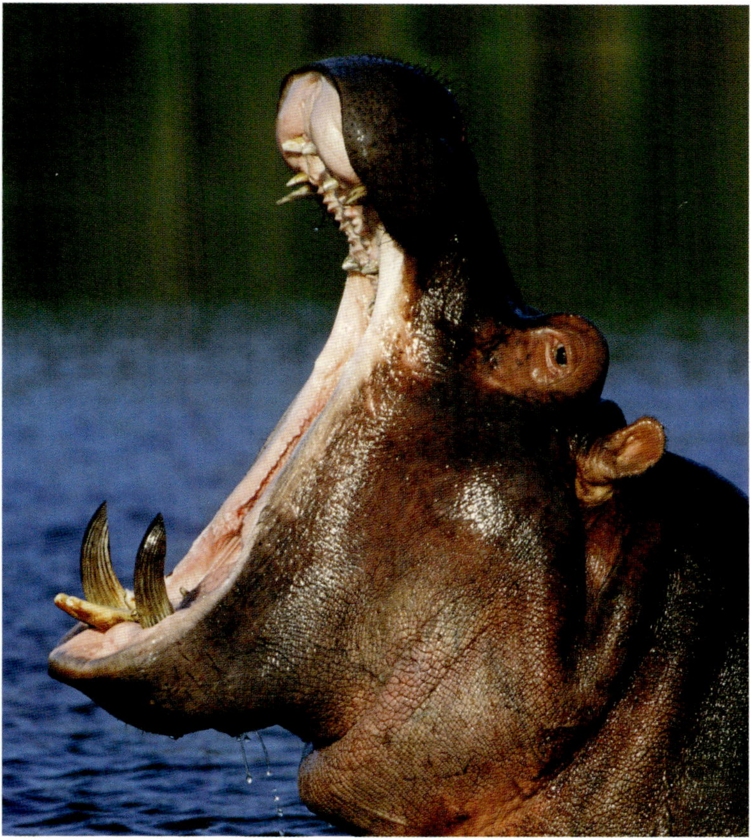

ALL PHOTOGRAPHS:

Hippopotamus

These social riverine animals are highly territorial. Males will often fight for dominance, using their large gape and pointed 'tusks' to inflict injury on a rival. Aggressive males sometimes target trespassing boats which can lead to capsizing. Hippos are thought to be responsible for up to 500 human deaths annually.

Lions attacking a buffalo
The 'King of Beasts' is a title well earned by this powerful and efficient predator. Here, a pride of females attacks a buffalo, which they will overpower and wrestle to the ground. The fatal bite will be administered to the neck, effectively suffocating the buffalo before the whole pride eats its fill.

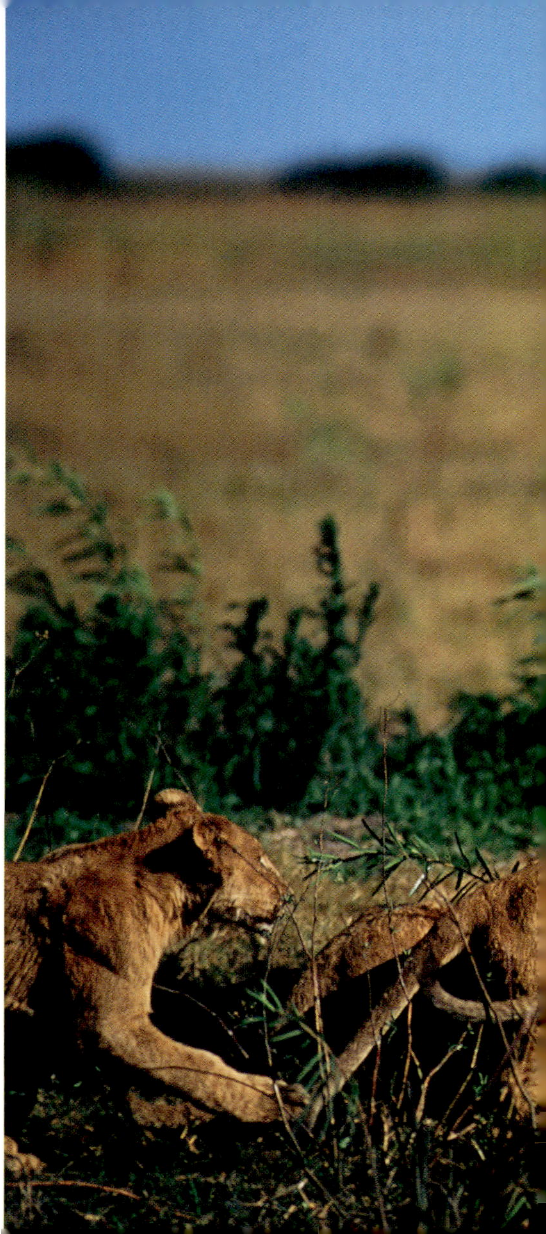

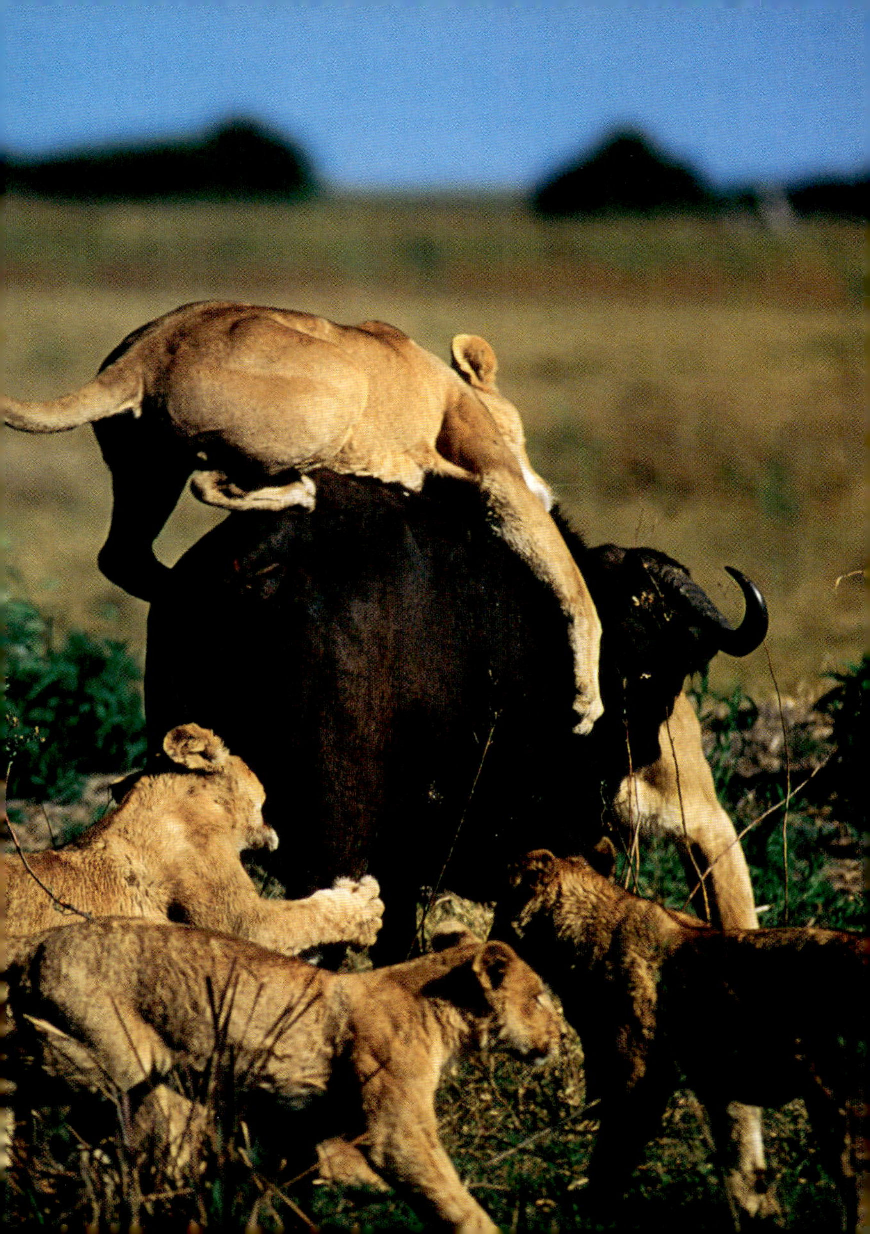

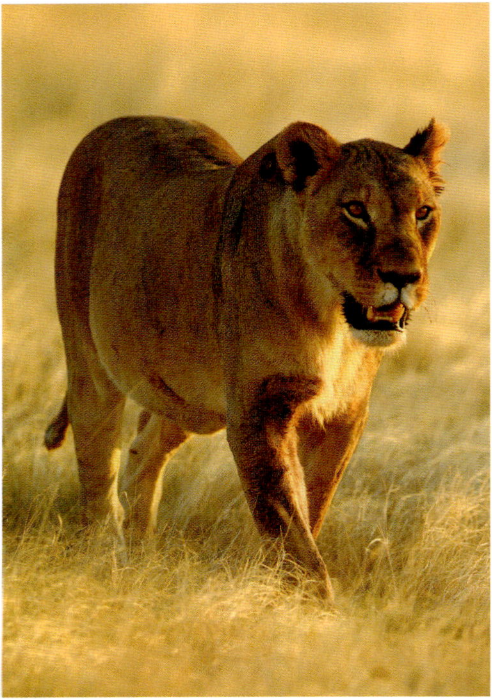

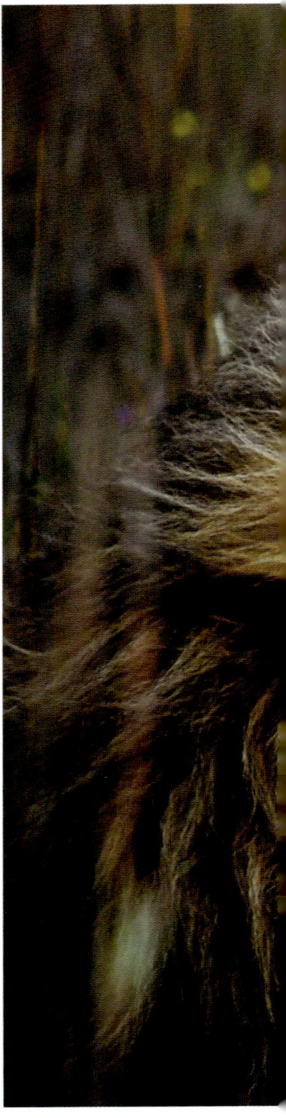

ALL PHOTOGRAPHS:

Lion

The concentration and teamwork of female lions is the key to hunting success, where stealth and a coordinated surprise attack are the primary weapons. The male here shows his long, golden mane and fearsome canine teeth, ideal for stripping flesh from the carcasses of unfortunate prey.

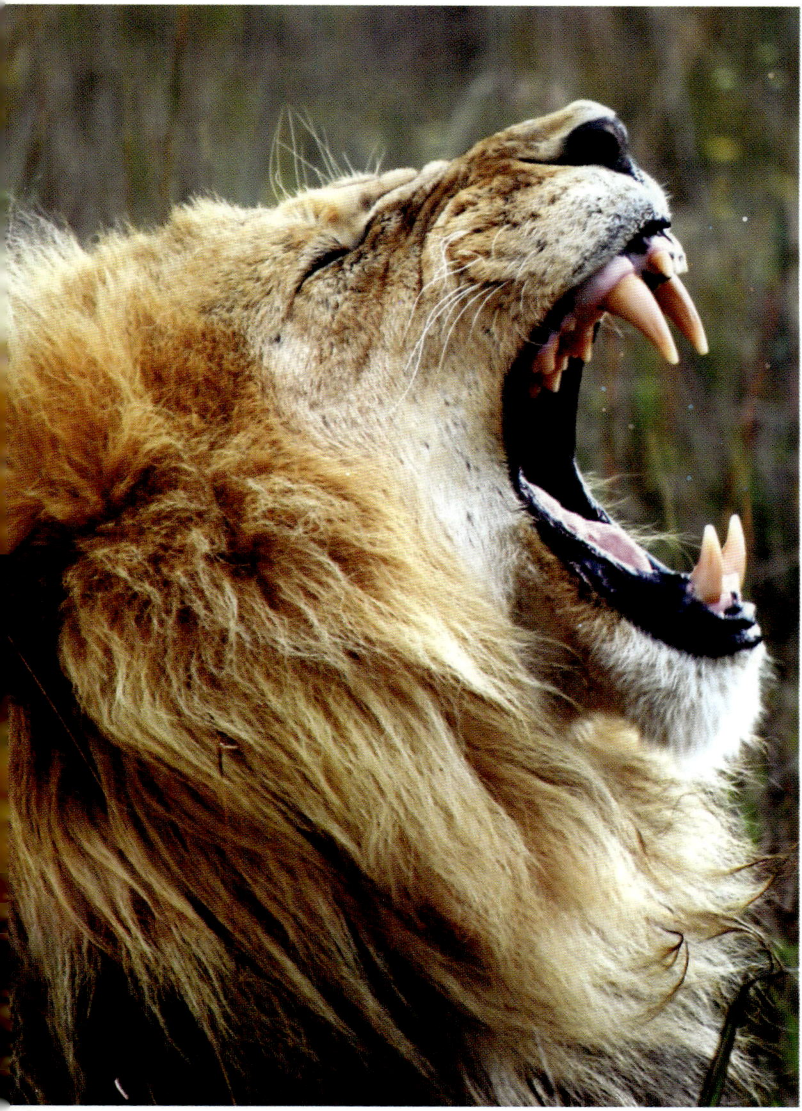

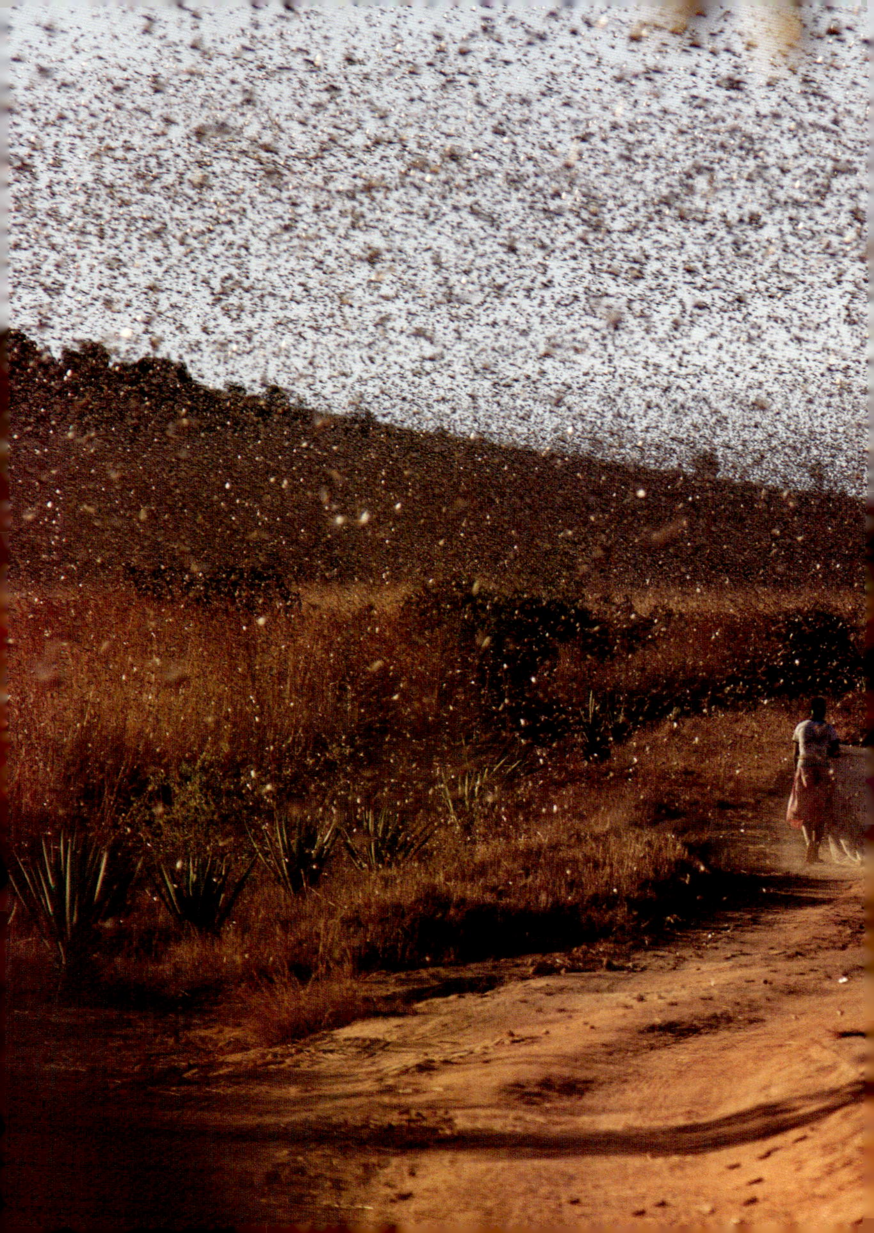

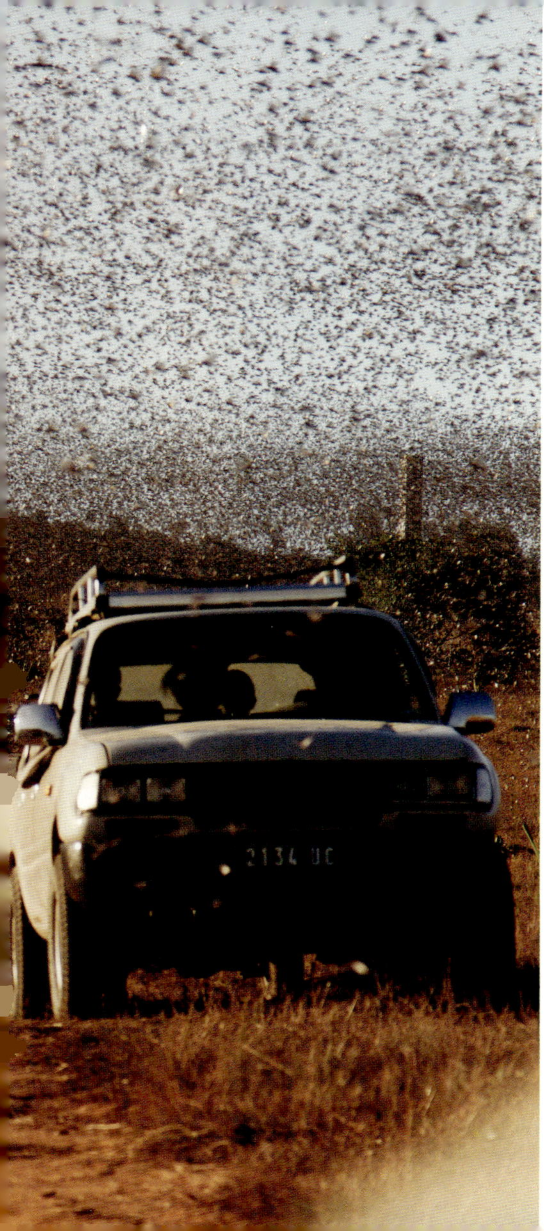

A swarm of locusts
When a period of drought is followed by rains and consequent rapid growth, the behaviour of some large grasshopper species changes. They breed abundantly and winged offspring form massive swarms that can ravage crops and cause widespread damage.

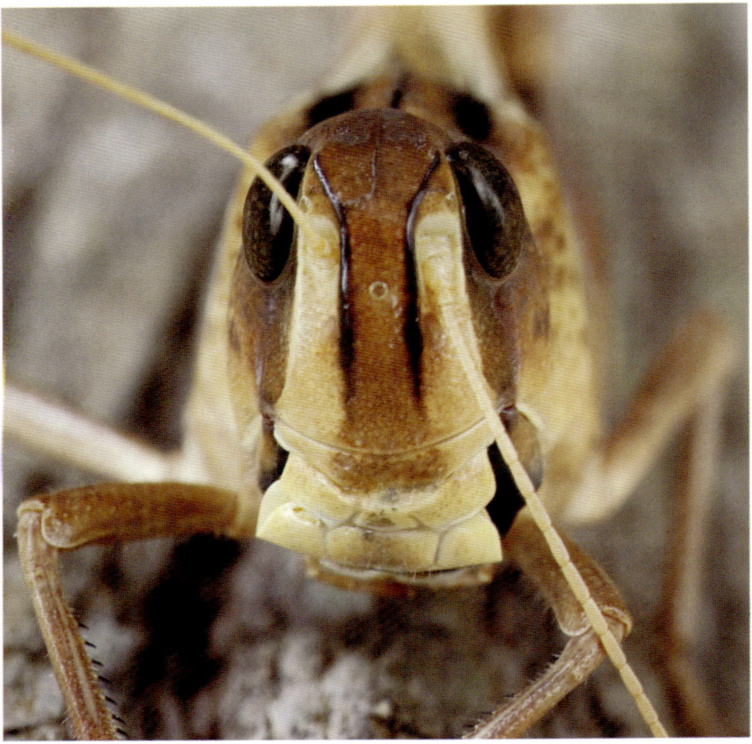

ABOVE:
Spur-throated locust
Locust swarms have featured in literature throughout history. Of particular note is the plague of Egypt mentioned in the Old Testament's Book of Exodus, a swarm that had devastating consequences for the population of the country.

OPPOSITE:
Black locust
A large locust swarm can comprise billions of insects spread out over a huge area of thousands of square kilometres. Population densities of these infestations are estimated to be up to 80 million insects per square kilometre (200 million per square mile).

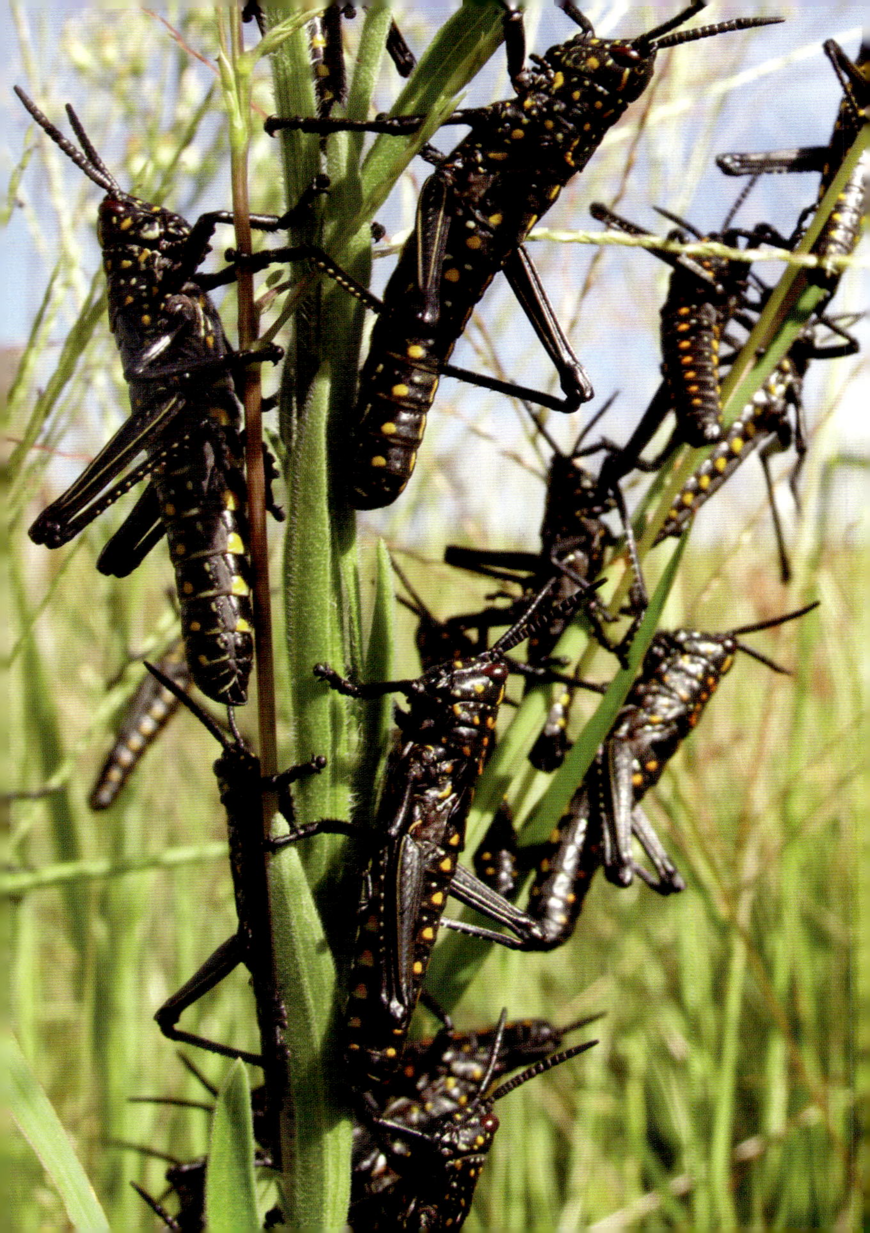

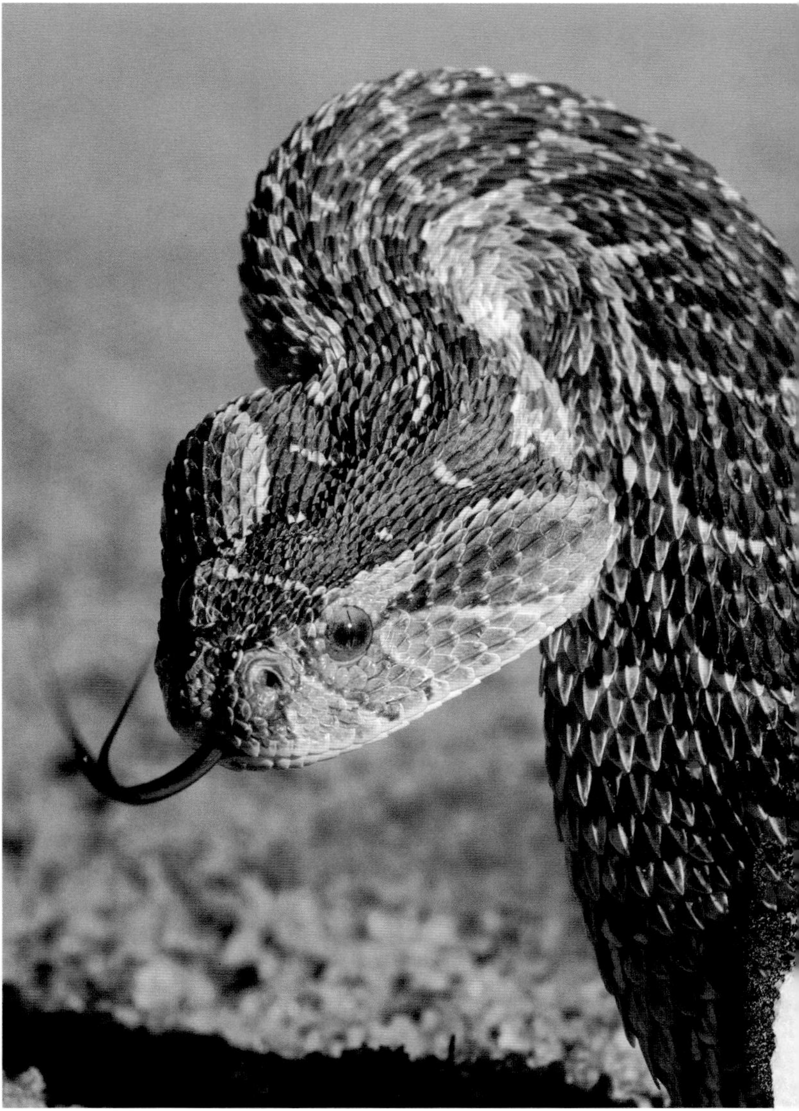

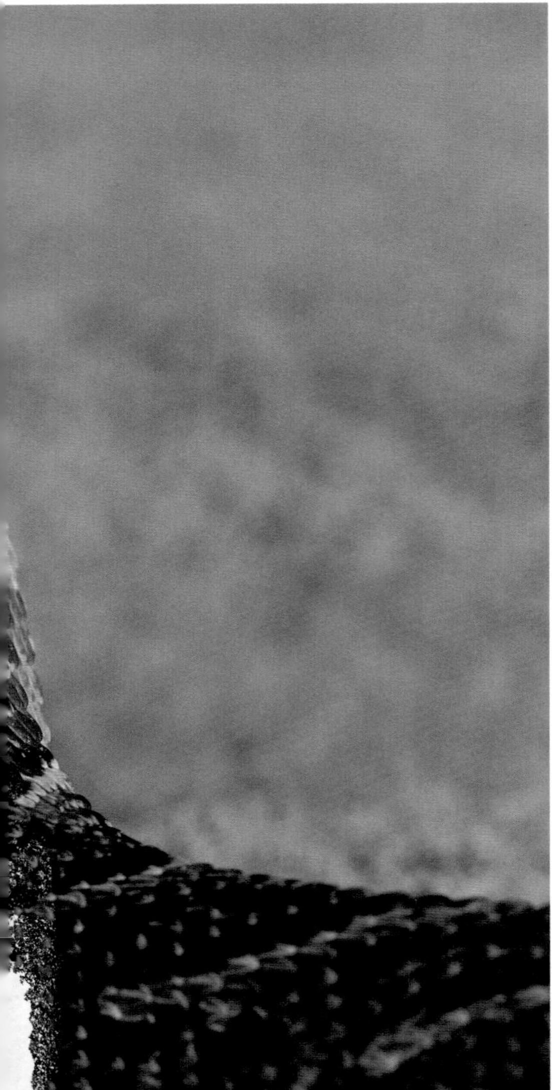

Puff adder

This highly venomous, 1m (3.3ft) long snake occurs throughout Africa's dryer regions. It is responsible for the majority of snakebite fatalities due to its mainly nocturnal habit of remaining motionless on paths in highly populated areas waiting for small mammals and lizards to stray close.

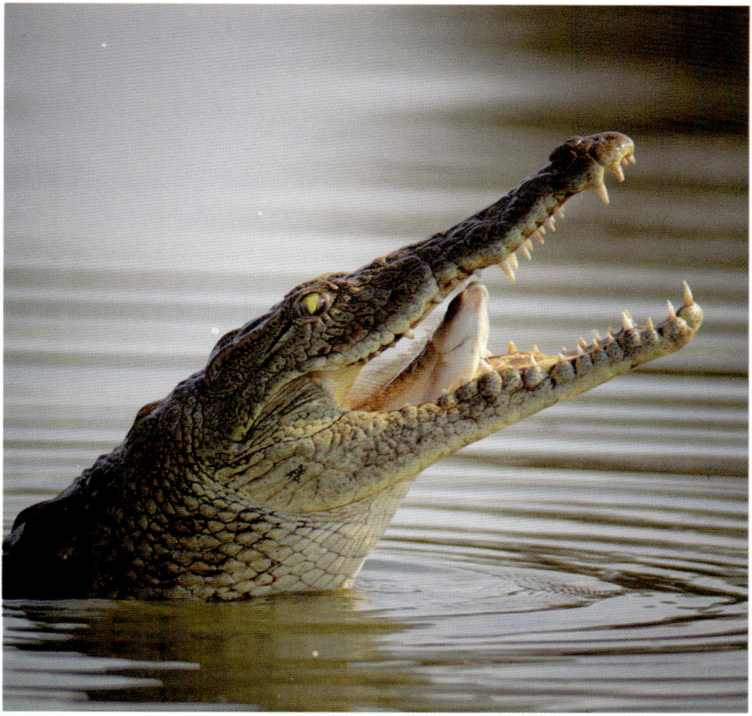

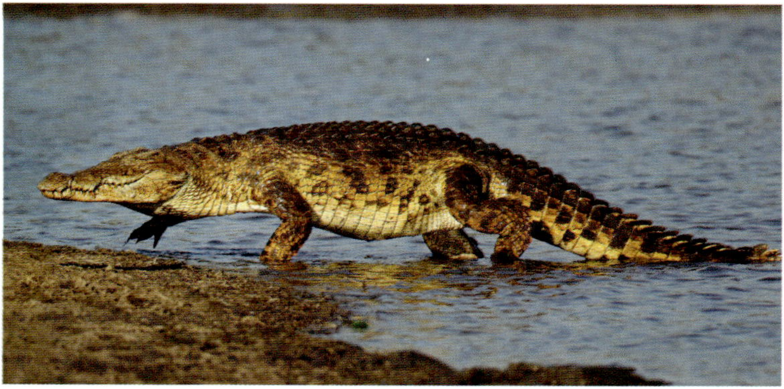

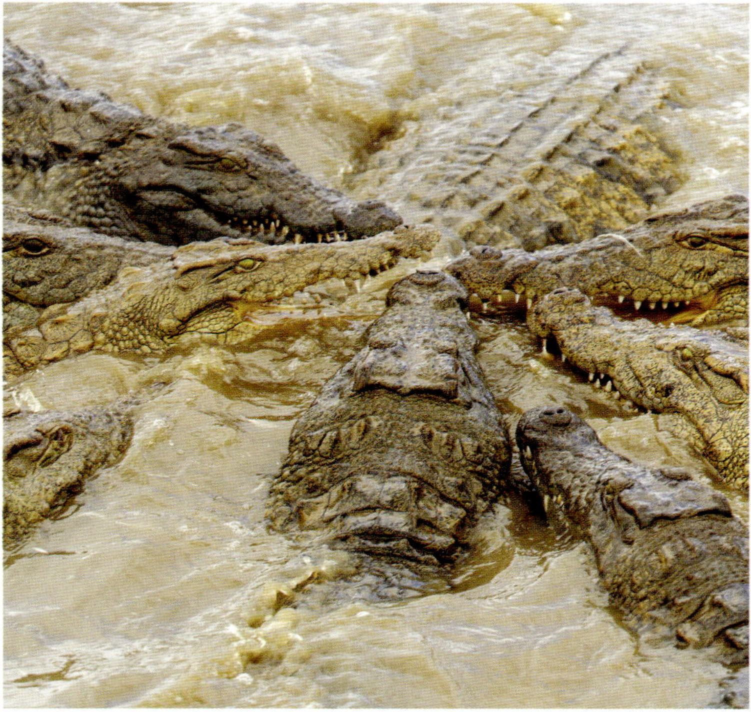

ABOVE AND OPPOSITE TOP:
Nile crocodile feeding
Nile crocodiles are apex predators, occurring in aquatic habitats across most of the African continent. Mature individuals mainly hunt by ambush, submerged in water for long periods of time patiently waiting for prey. Mass feeding on large carcasses is a common habit. Crocodiles have a diet of fish, carrion and just about any live animal they can catch. A crocodile has successfully caught a large fish which will be swallowed whole. The powerful jaws are used to grip on to prey rather than chewing flesh.

OPPOSITE BOTTOM:
Nile crocodile
Crocodiles are relatively clumsy on land. Males average a length 3.5–5m (11–16.5ft) and weigh up to 750kg (1,650lb). Females are approximately 30 per cent smaller. It is thought that Nile crocodiles are responsible for several hundred human fatalities a year.

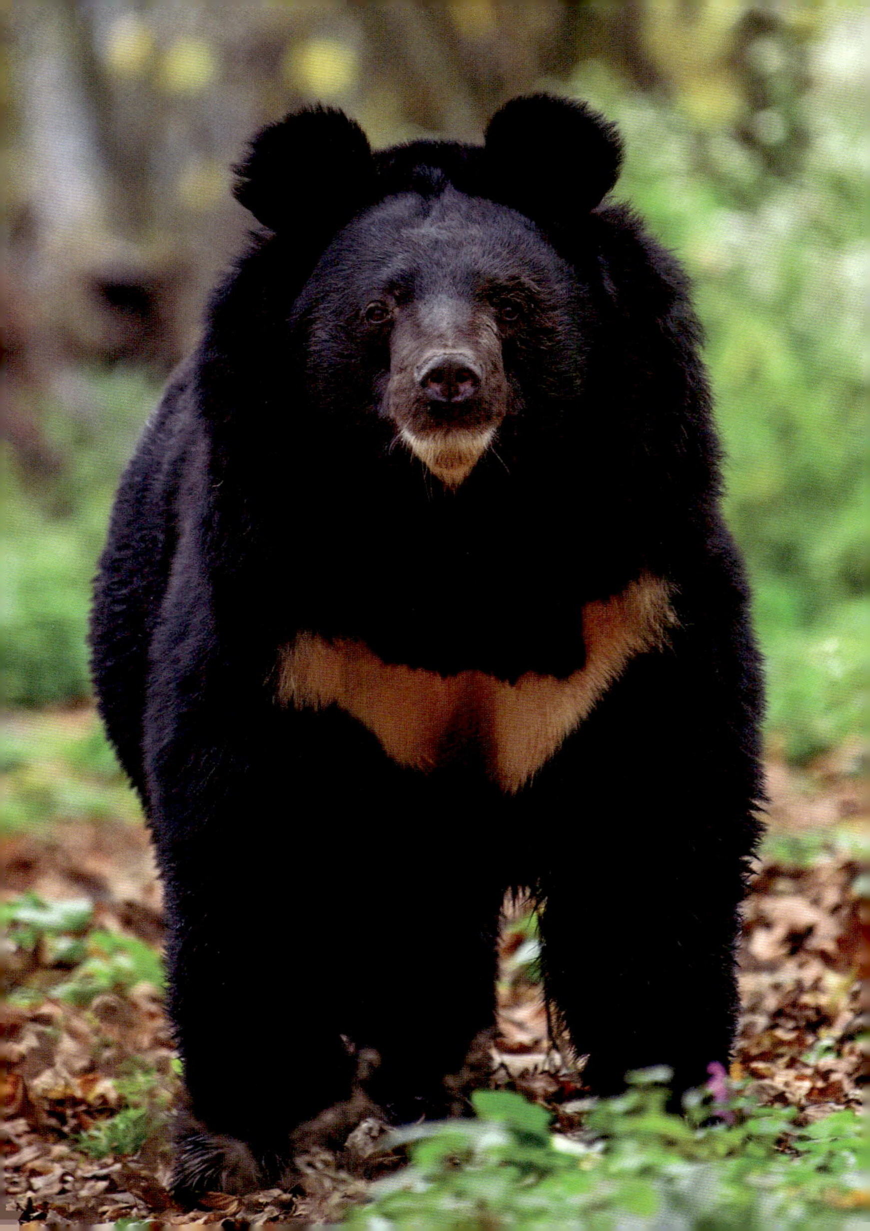

Asia

Asia is the name given to a wide band of countries that spread from the borders of Europe in the west to the borders of the Pacific Ocean in the east. It is collectively the largest continent, covering an area of 44 million sq km (17 million sq miles). It is also the most populous continent on Earth, with some 4.5 billion people calling it home. Contrast this with the knowledge that the world population of all six subspecies of wild tigers is a little under 4,000, spread across just 13 Asian countries.

Despite its high population, Asia has within its boundaries some of the world's greatest natural wonders: the sheer breath-taking beauty of the Himalayas with many of the world's highest peaks, the world's deepest cave at Son Doong in Vietnam, the vastness of the cold Gobi Desert, and lush islands surrounded by soft white sands and shallow seas full of exotic marine life. Not surprisingly, there is a great diversity of animals inhabiting the area, some of which are endemic to a country or particular climatic zone and are found nowhere else on the planet. It hosts the smallest bat on Earth, the 3cm (1.2in) long Kitti's hog-nosed (or bumblebee) bat, native to Thailand and Burma, and the largest lizard, the impressive Komodo dragon featured later in this chapter.

The Himalayas and other mountainous regions of central Asia hold the total world population of the enigmatic snow leopard, whilst the world's longest venomous snake, the king cobra, can be encountered in India and other south-eastern countries.

OPPOSITE:

Asiatic black bear

A medium-sized bear with a distinctive V-shaped white patch on the chest, the Asiatic black bear has a scattered and vulnerable population, ranging through the Himalayan region, northern India, Korea, and parts of China, Russia and Japan.

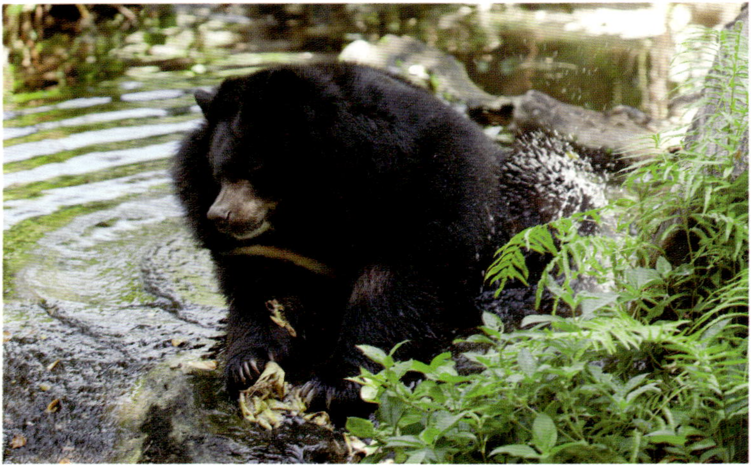

ALL PHOTOGRAPHS:
Asiatic black bear
These animals are superbly adapted to life in forests. They are omnivorous, with a wide-ranging diet including honey, fruit, invertebrates and seeds. They are a protected species throughout their range, but sometimes come into conflict with commercial farming and are consequently persecuted in some remote regions.

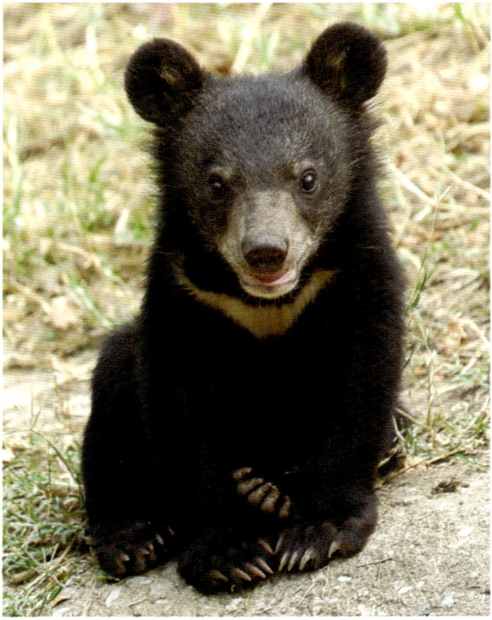

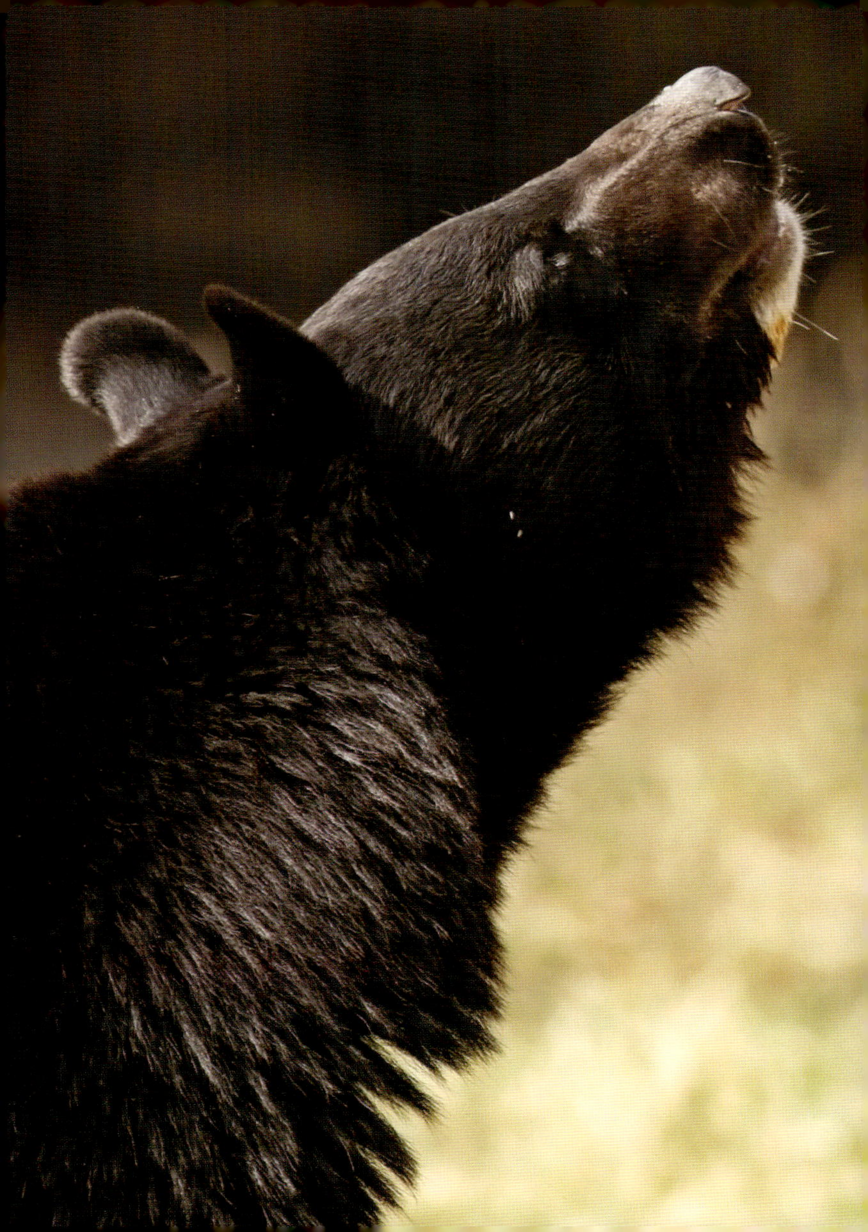

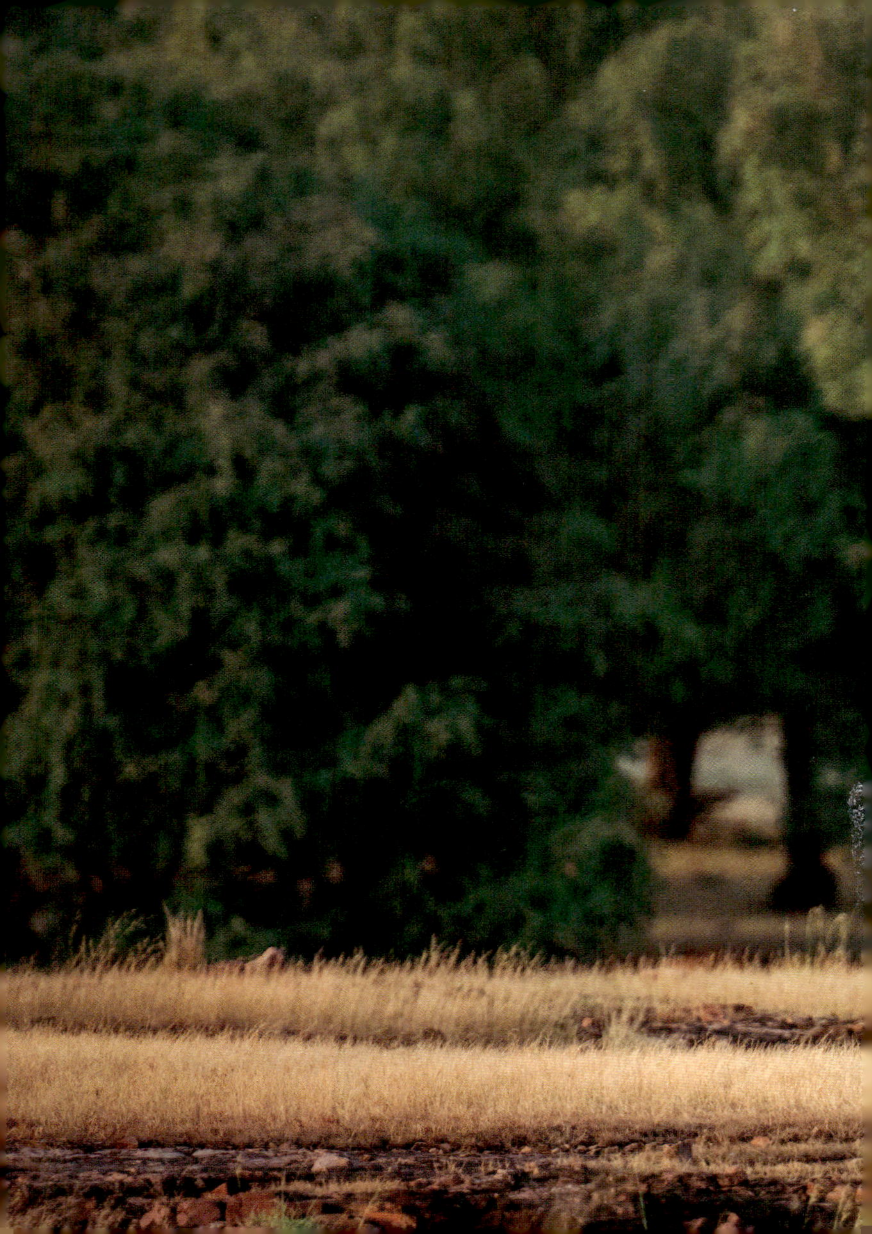

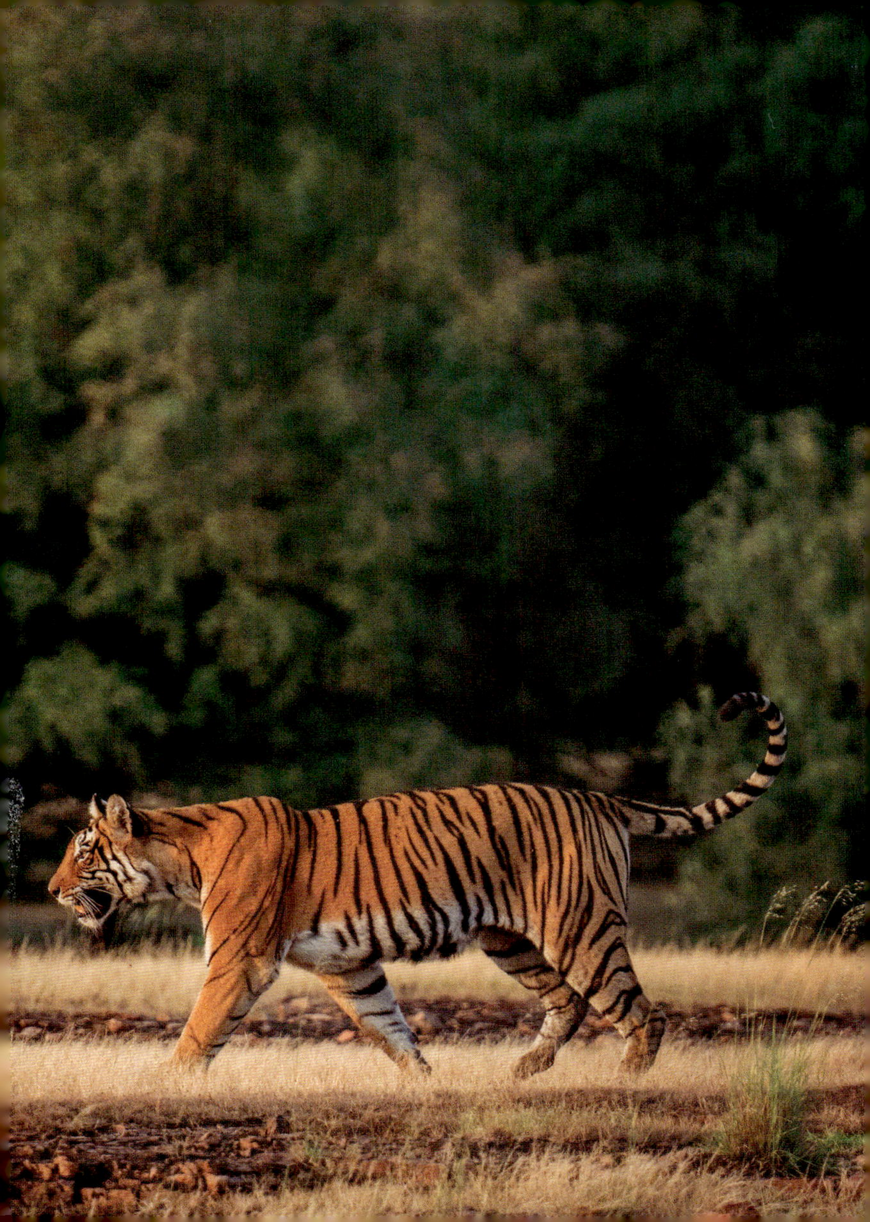

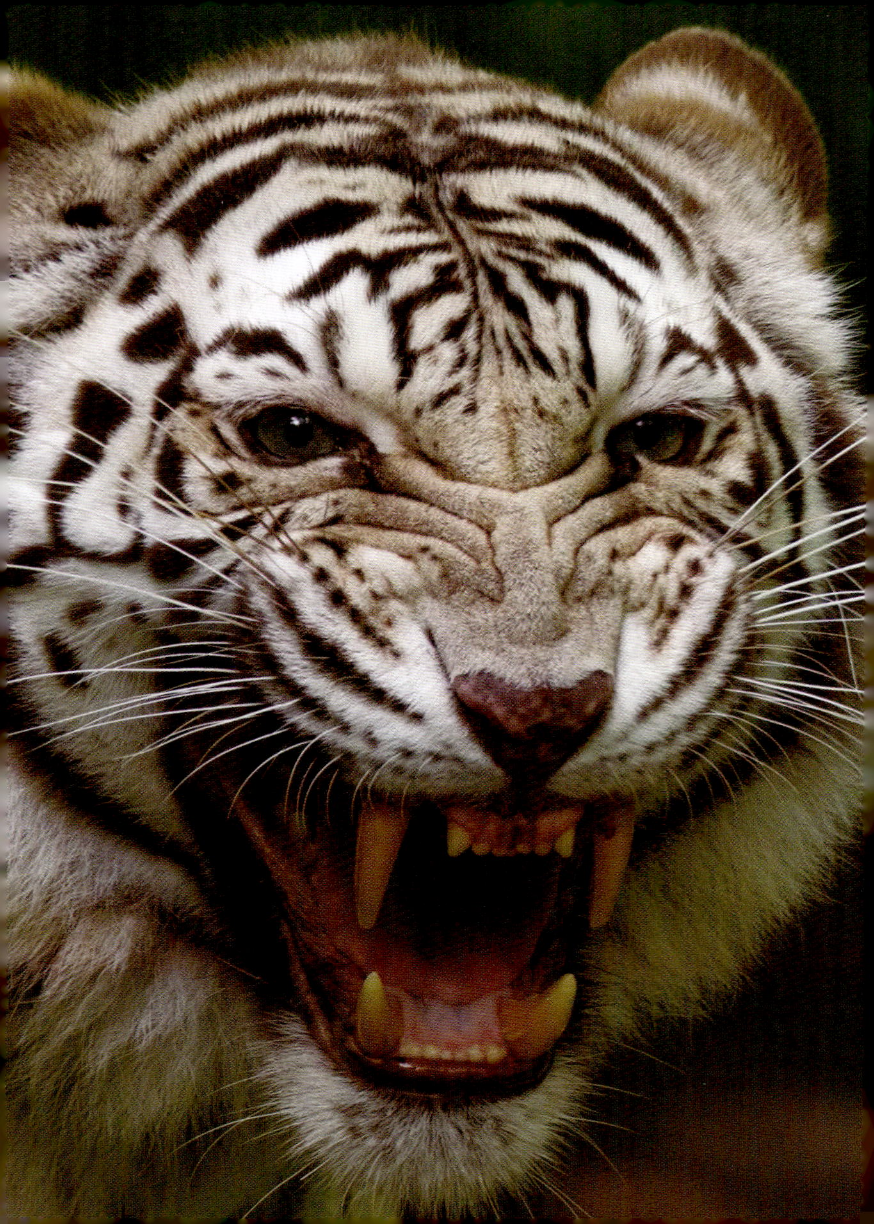

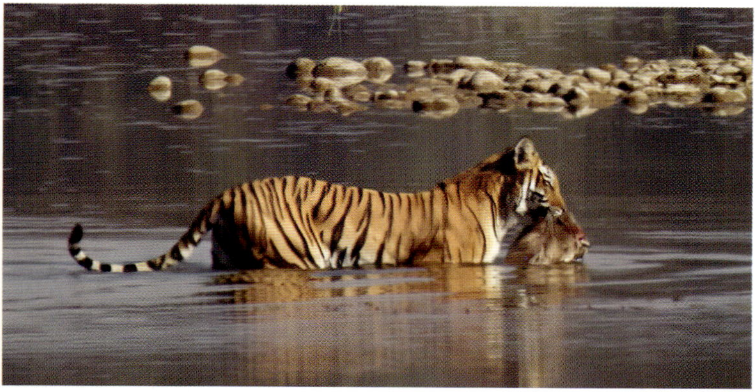

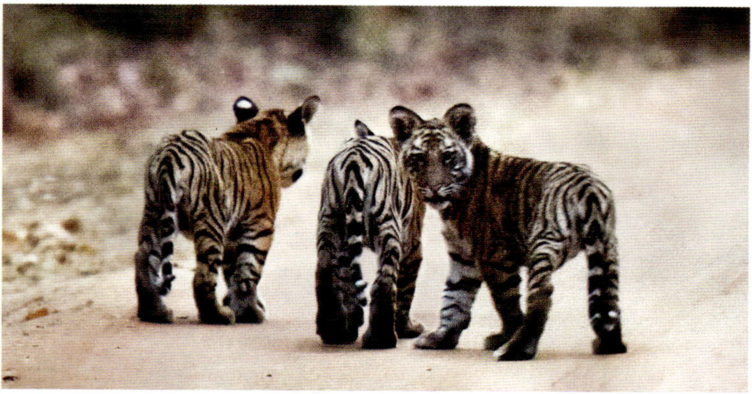

ALL PHOTOGRAPHS AND PREVIOUS PAGES:

Bengal tiger

The sight of a majestic tiger striding confidently through its native habitat (previous pages) is one of the world's most stirring animal encounters. Bengal tigers are native to the Indian subcontinent where it is estimated only some 1,500 are now living in the wild. The sub-species is under extreme threat from poaching and habitat loss. These are strong animals with exceptional hearing and eyesight. They have the largest canines of any big cat, measuring some 10cm (4in). They hunt deer and other animals by stalking, when their orange colouration, broken by a series of black lines, acts as perfect camouflage.

Box jellyfish

There are over 50 species of box jellyfish occurring across the world's oceans. Some species are highly venomous, with their sting causing severe pain and even death to humans. The name derives from the square, box-like bell with a tentacle on each corner. These tentacles can reach a length of 3m (10ft).

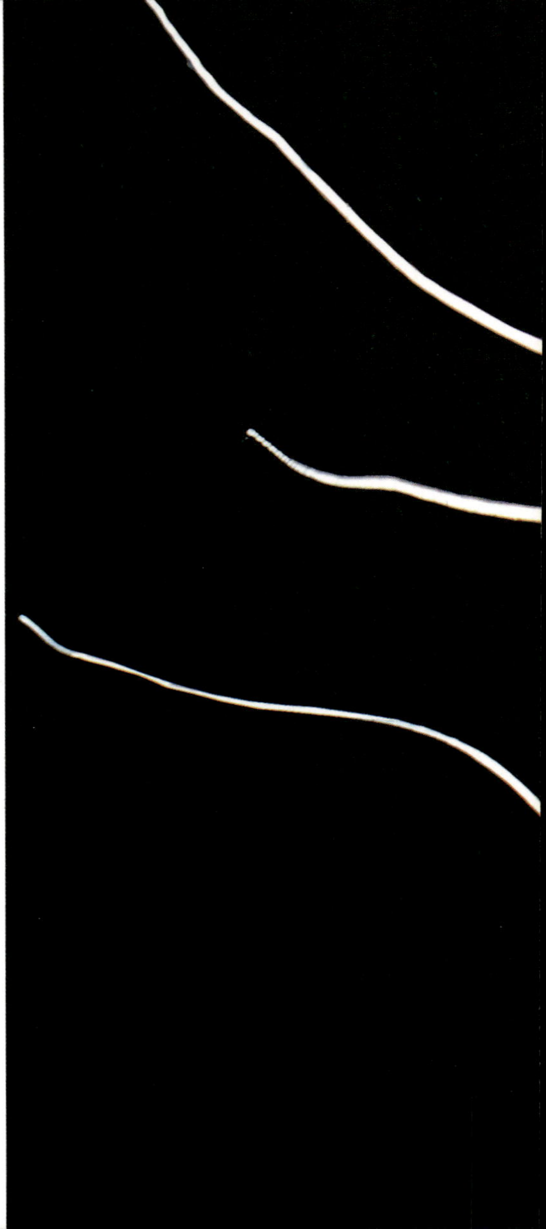

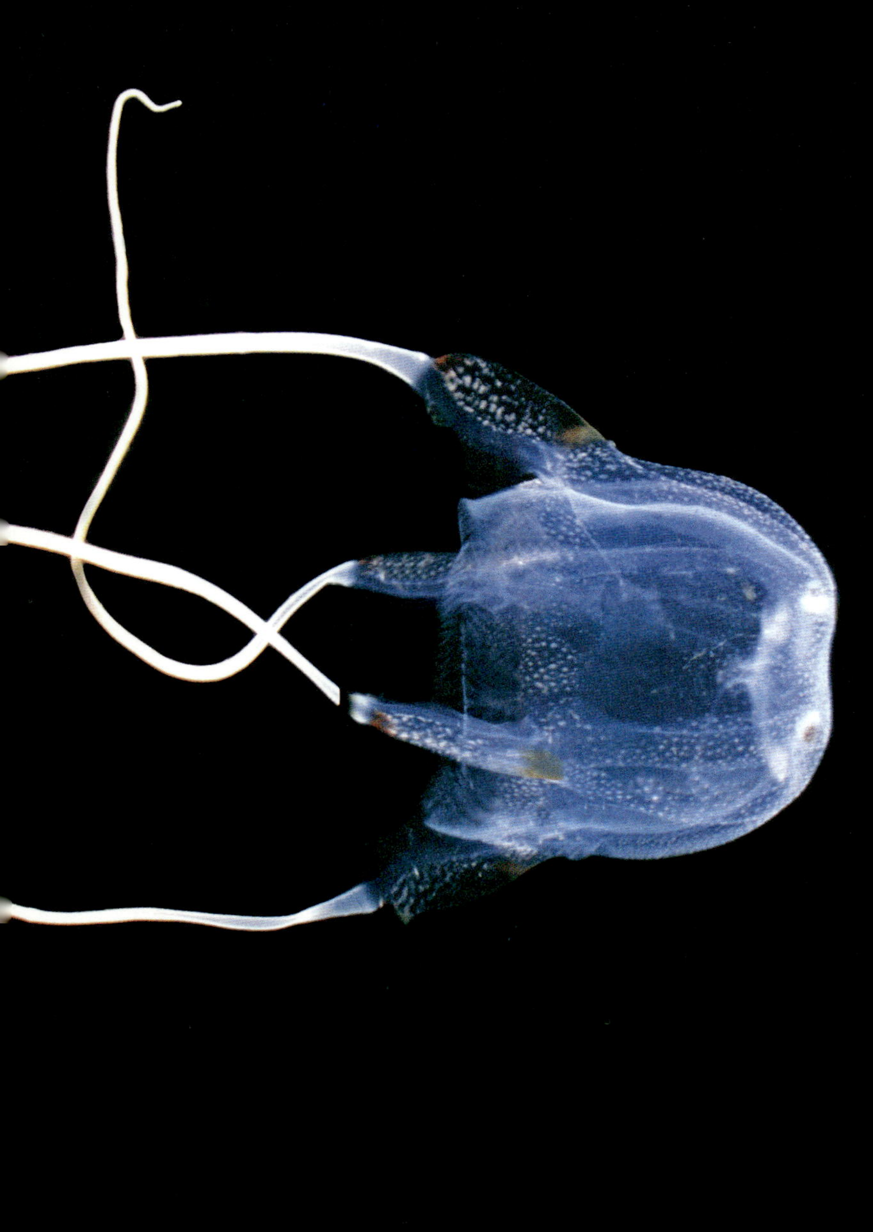

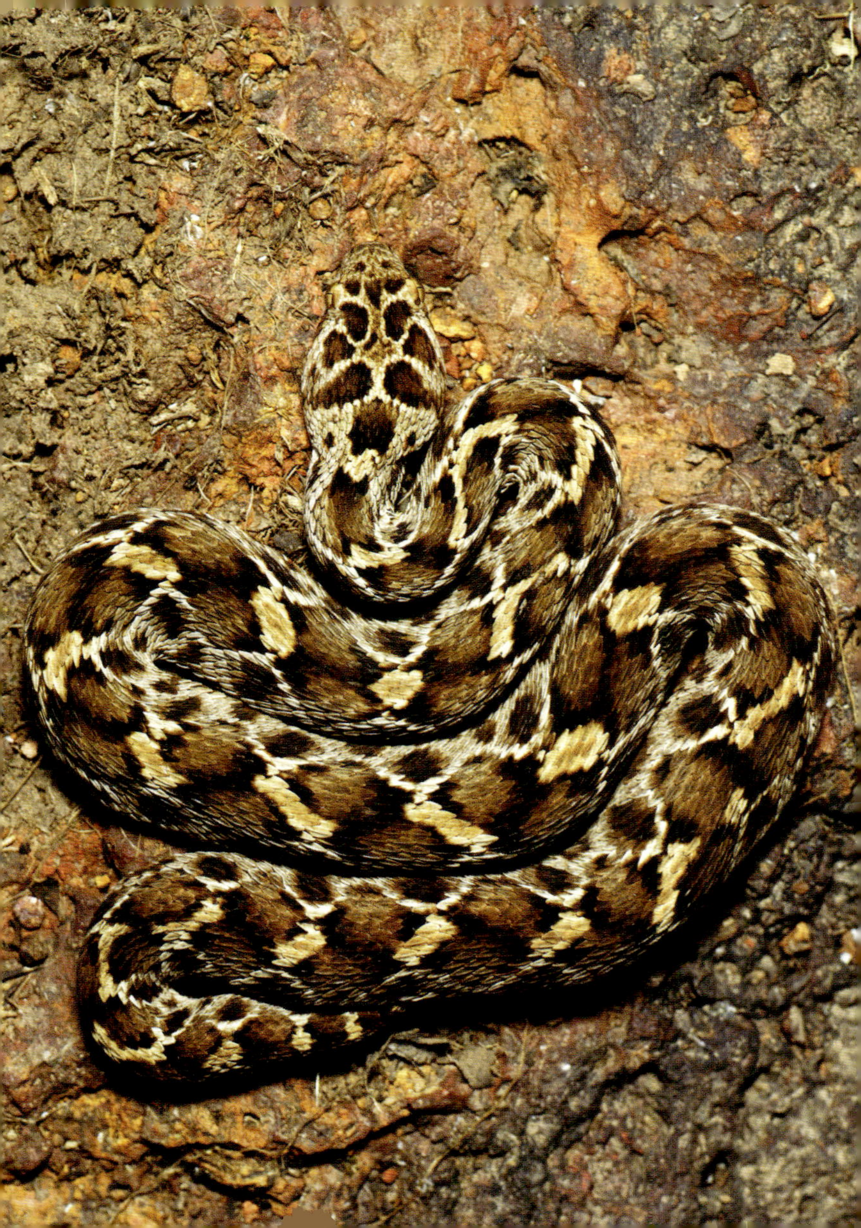

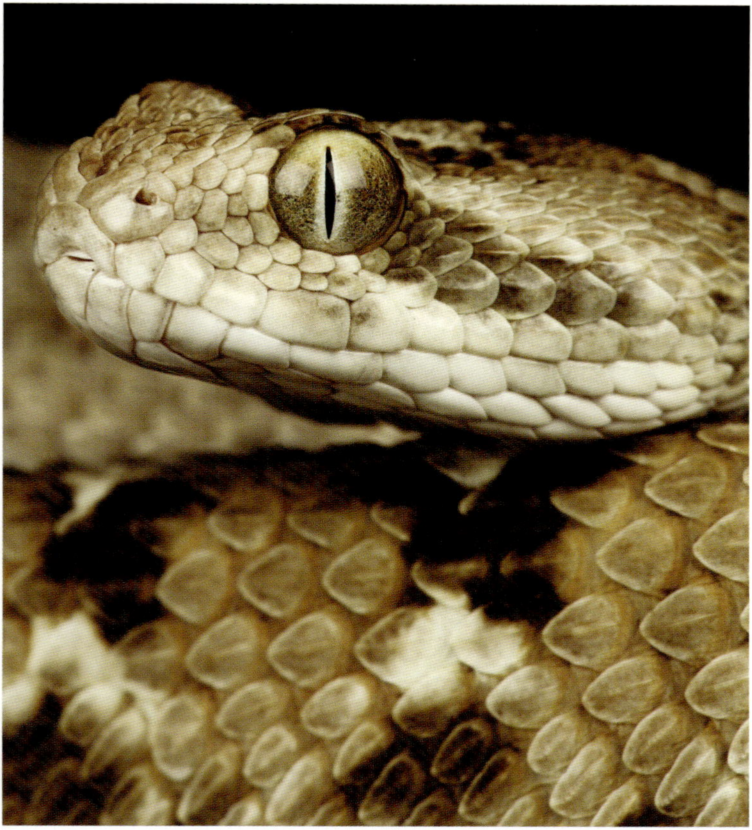

ALL PHOTOGRAPHS:

Indian saw-scaled viper

So-called after the serrated scales on its flanks, this snake grows to an average length of 60cm (24in). Its repuscular habits, aggressive nature and excellent camouflage make it one of the most dangerous snakes in the region. However, nowadays, human deaths are rare due to effective and widely available antivenom.

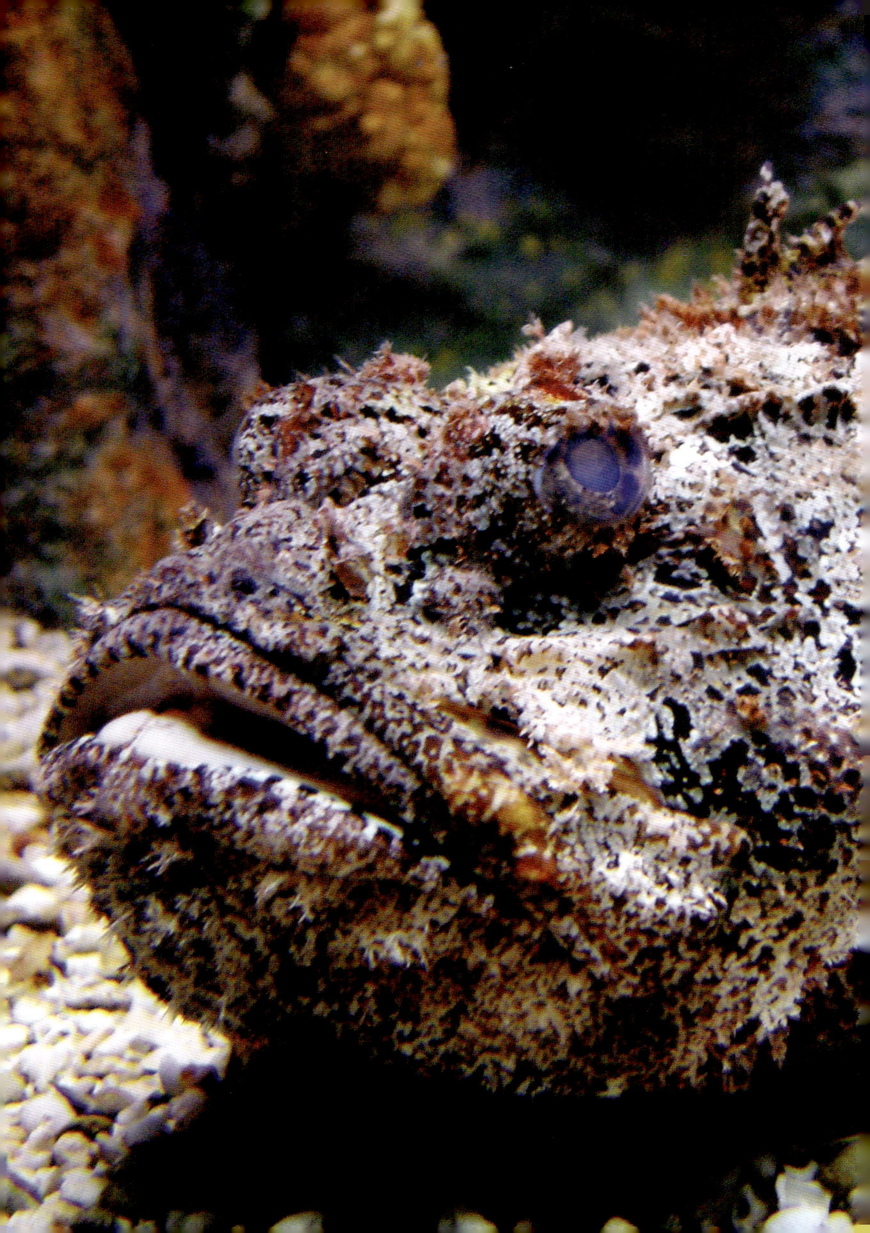

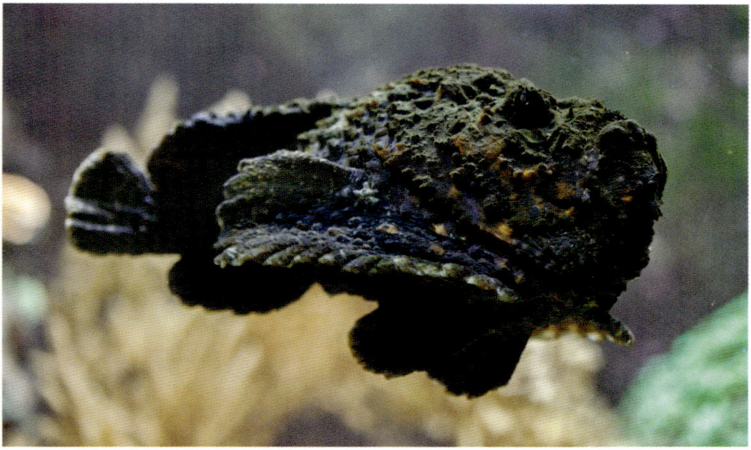

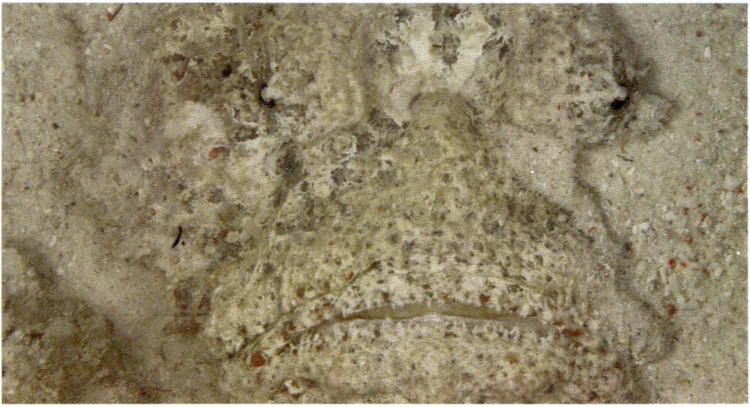

ALL PHOTOGRAPHS:

Stonefish

This well camouflaged fish lives in coral reefs of tropical Indo-Pacific oceans. If a person inadvertently treads on the fish, the dorsal spines raise and inject a powerful venom into the victim. This sting is extremely painful and, if left untreated, can be fatal.

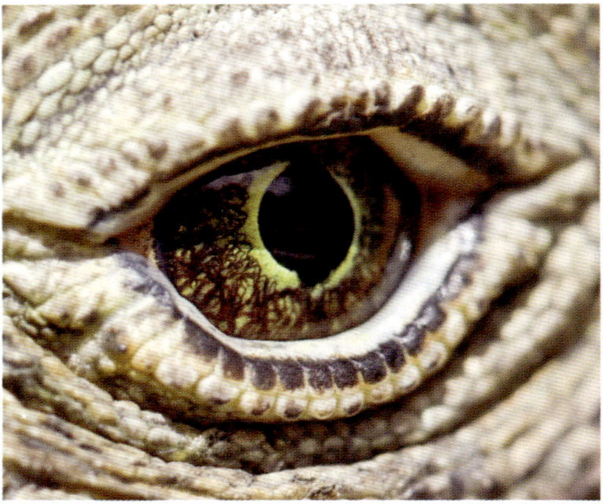

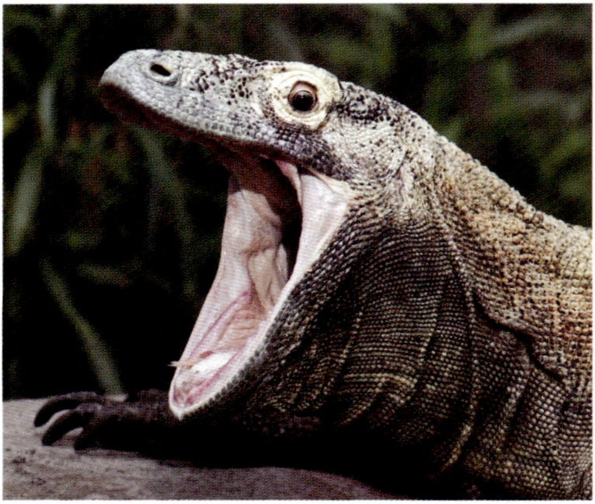

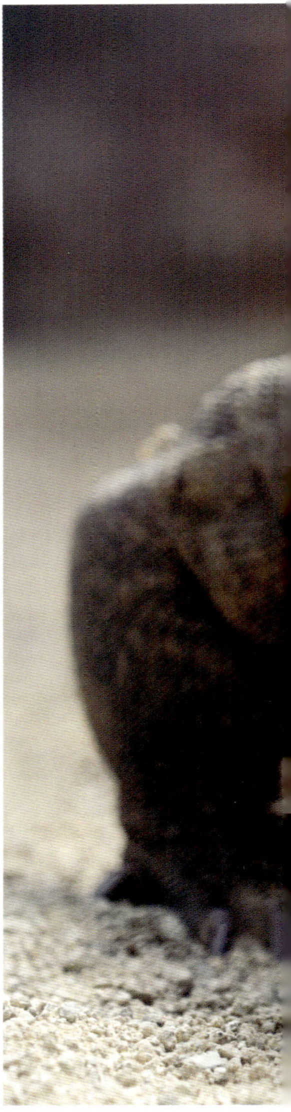

ALL PHOTOGRAPHS:

Komodo dragon

Endemic to a few Indonesian islands, this large lizard, reaching up to 3m (10ft) long, is an apex predator. It hunts by ambushing a wide range of live prey, but will also feed on carrion. It has an unwarranted fearsome reputation, since attacks on humans are actually very rare.

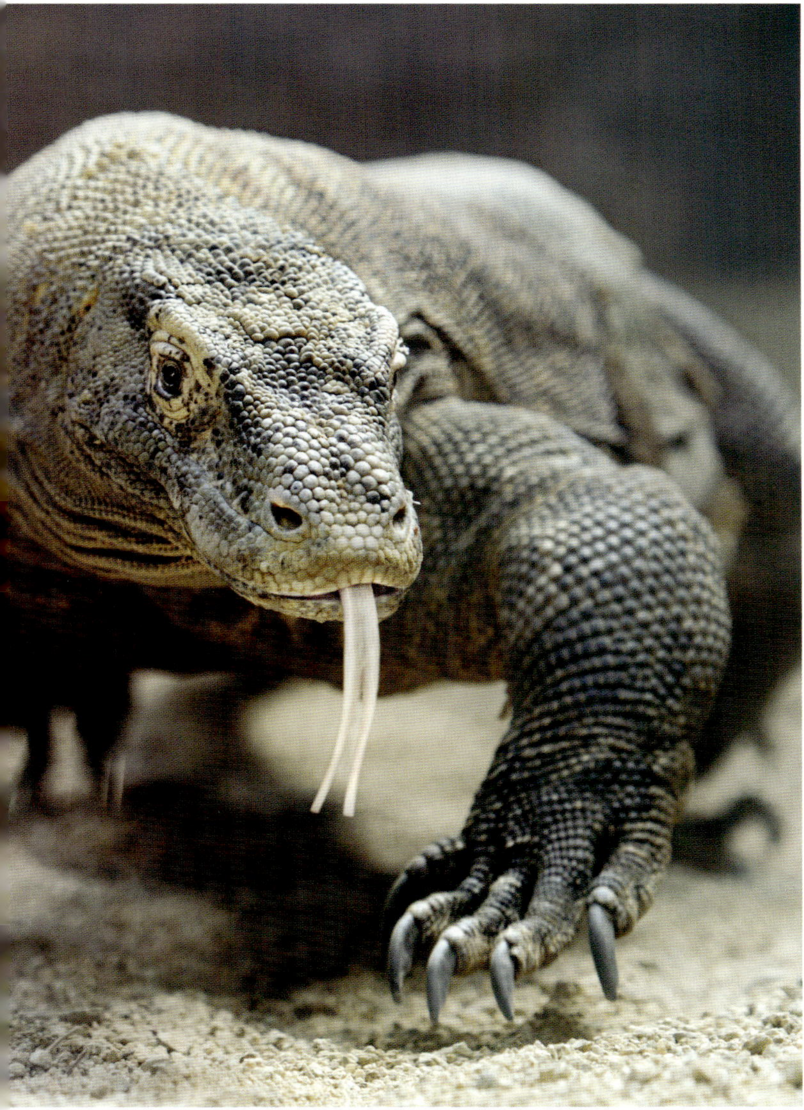

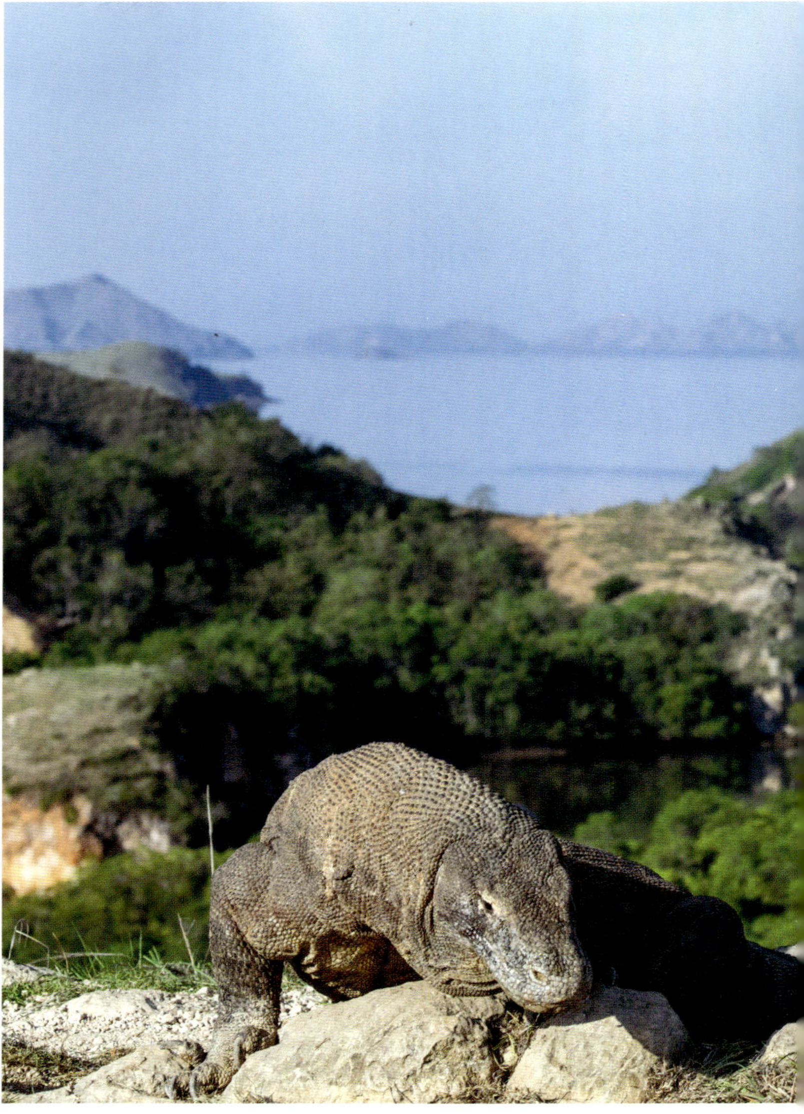

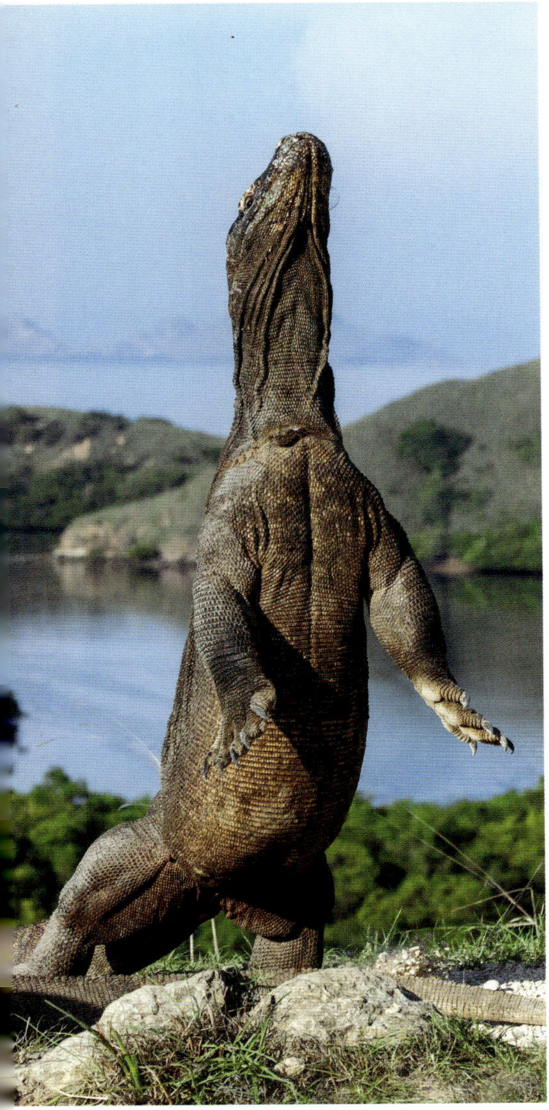

Komodo dragon
A threatened species, the current population is estimated to be approximately 3,000 individuals. Poaching of the animals themselves, and the prey on which they depend, is a major factor in their decline. Climate change, with its potential for degradation of essential habitats, may lead to further losses.

ALL PHOTOGRAPHS:
King cobra
The world's longest venomous snake, reaching a length of 4m (13ft), the king cobra can live for up to 20 years. Its primary prey are other snakes, which it hunts by 'smelling' them with its forked tongue. King cobras avoid humans, but if surprised will raise their bodies and extend its 'hood' as a means of defence.

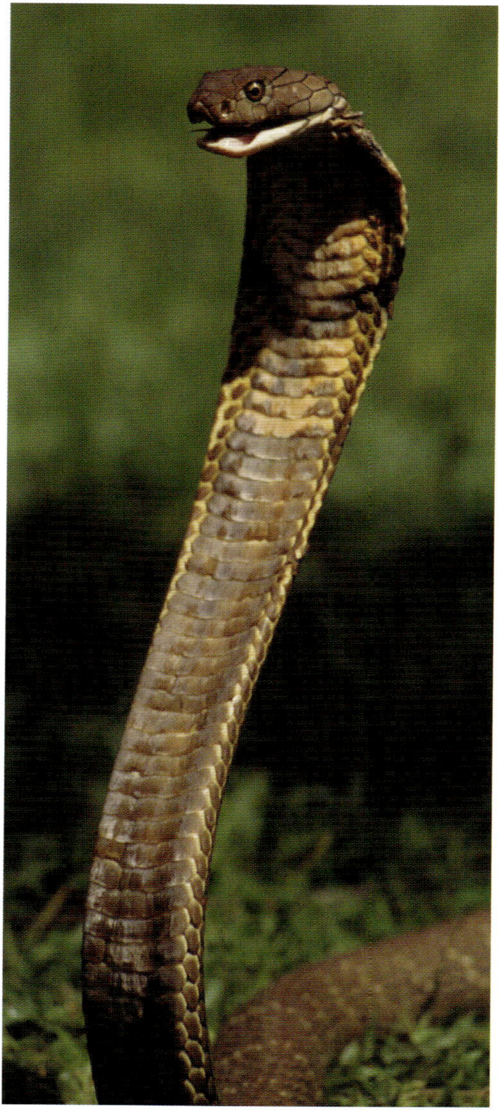

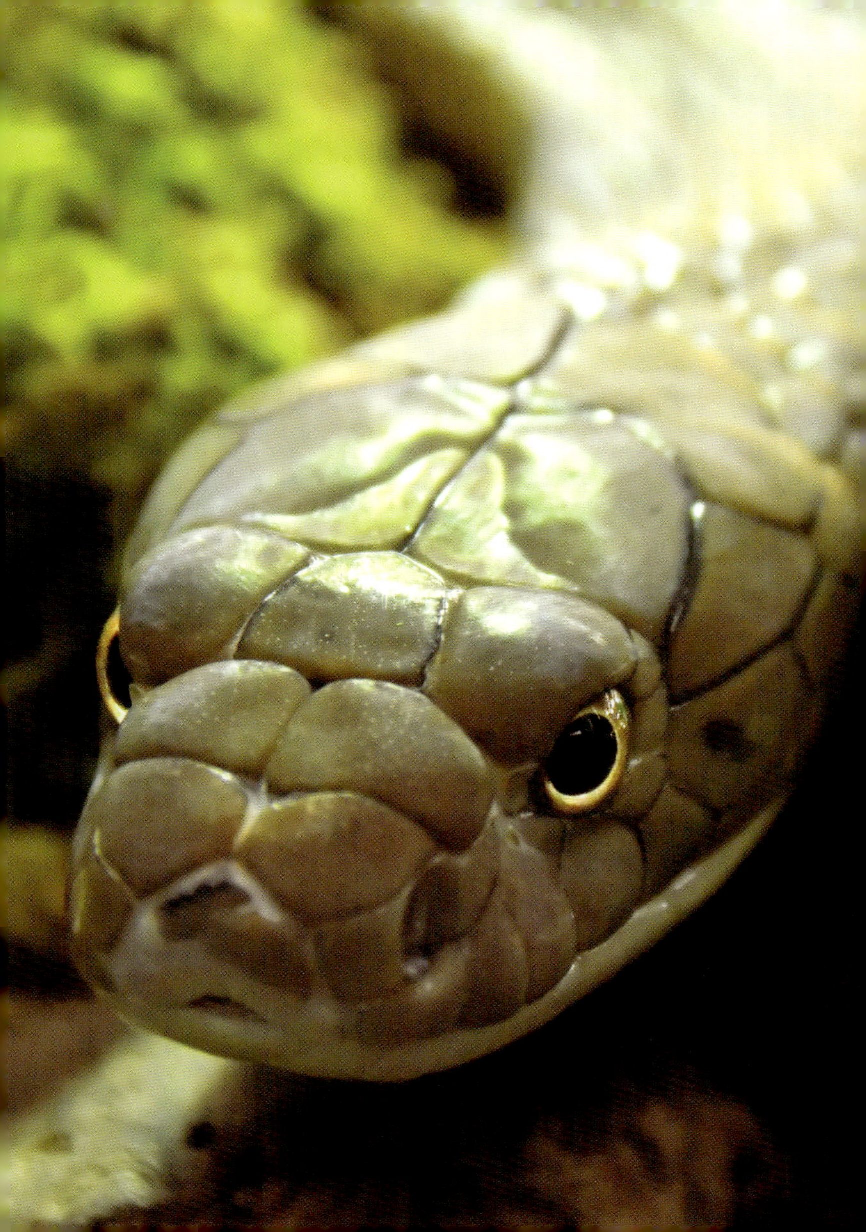

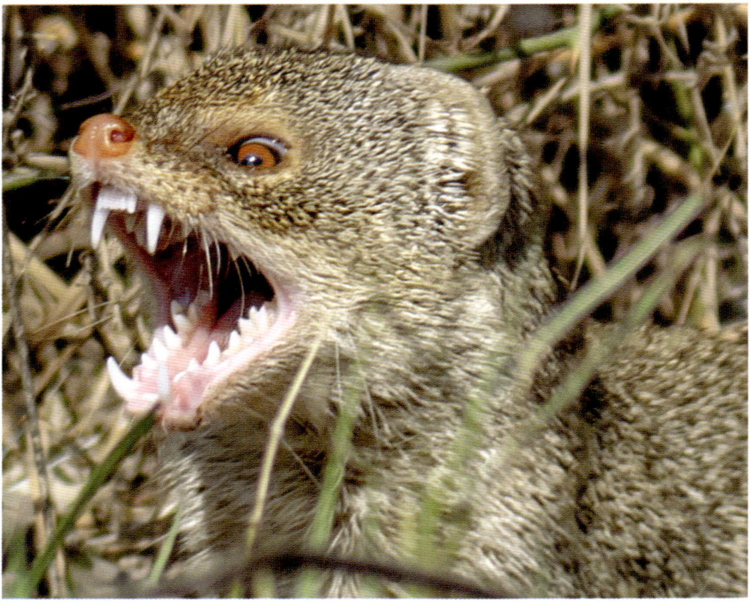

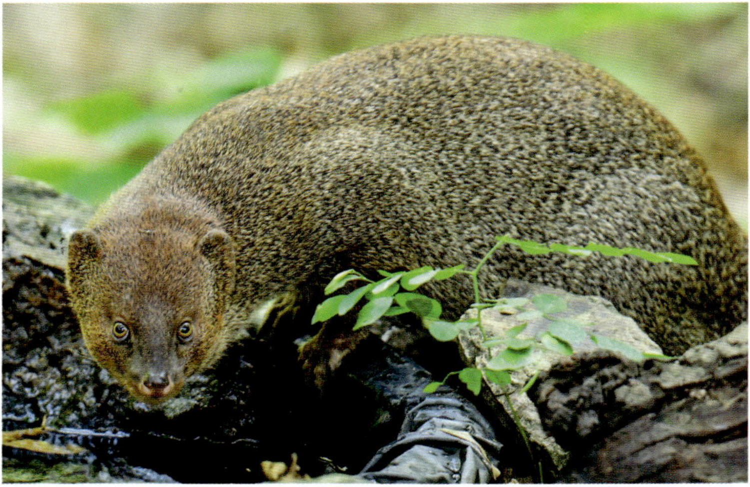

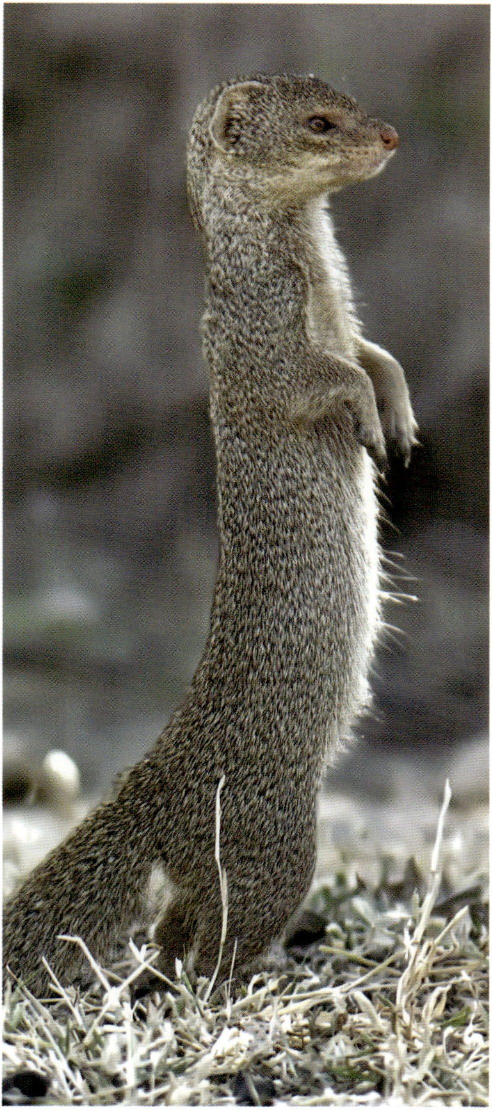

ALL PHOTOGRAPHS:

Indian grey mongoose

Native to the Indian subcontinent, this agile, inquisitive and legally protected species feeds on snakes, scorpions, rodents and invertebrates. It is so called because the grey fur is much stiffer than in other species. It is commonly kept as a pet in order to keep dwellings free of rats and other pests.

Snow leopard
This beautiful and endangered species inhabits high mountainous areas of central and southern Asia. There are estimated to be fewer than 10,000 individuals left in the wild, with poaching and the illegal fur trade contributing to ongoing decline. It feeds primarily on wild sheep and goats.

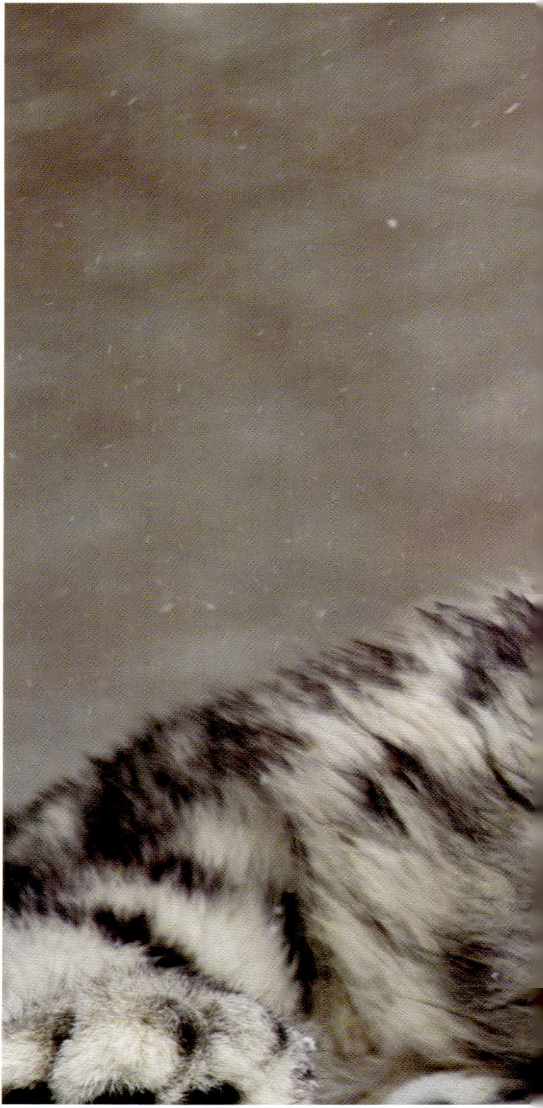

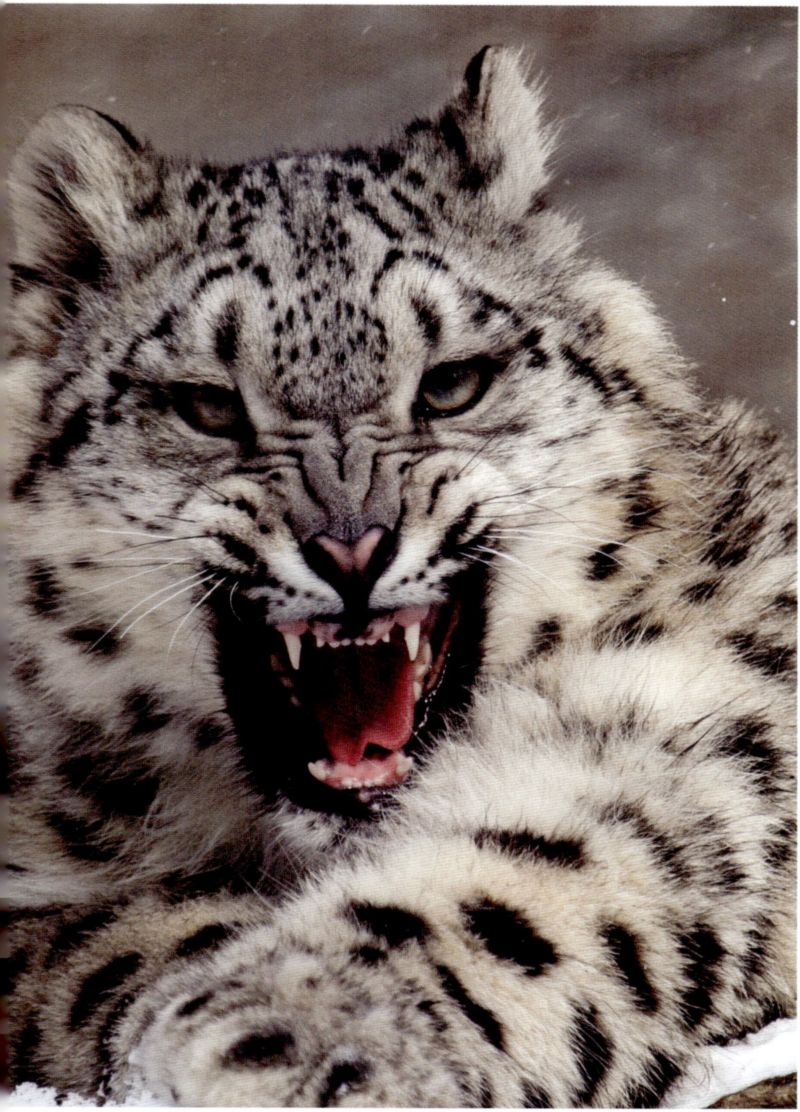

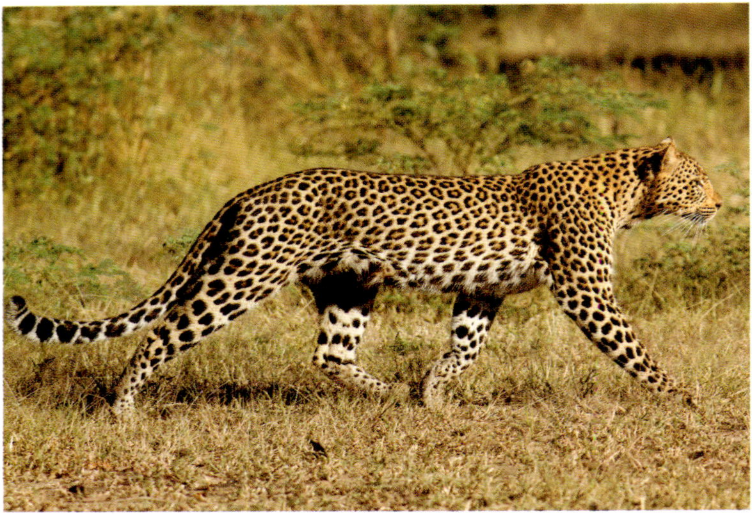

ALL PHOTOGRAPHS:

Leopard

Widely distributed in the region, with seven recognised sub-species, this powerful predator is at home in forested habitats. It is a stealthy predator, easily able to stalk its preferred prey of deer, primates and other medium sized animals to close distance. Although attacks on humans are rare, some fatalities are recorded every year.

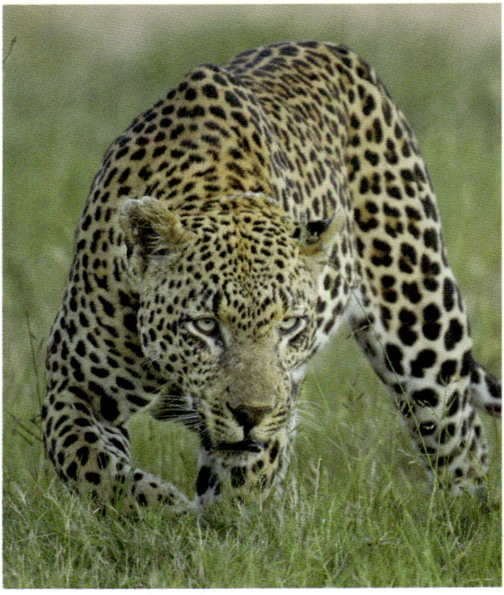

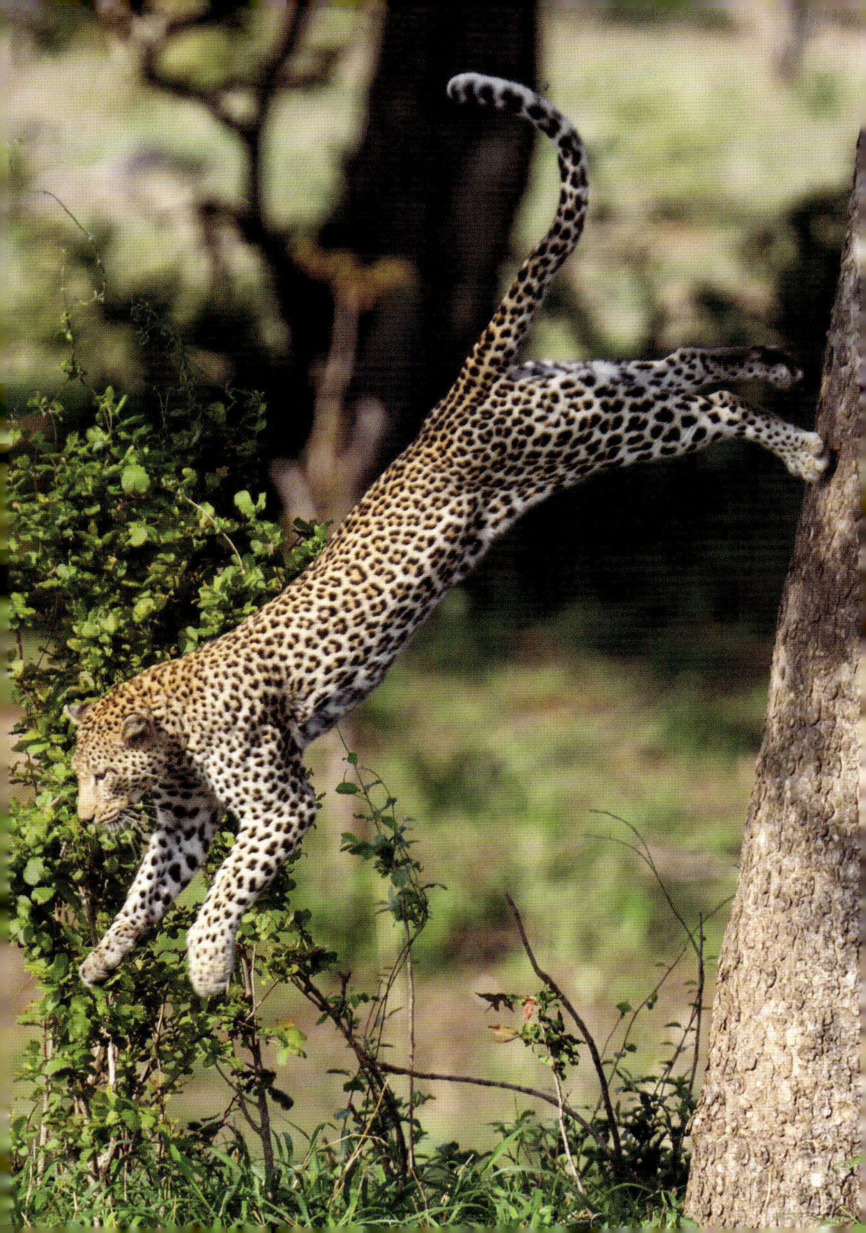

Reticulated python
'Reticulated' means 'net-like', a reference to the intricate patterns on this snake's skin. It is the world's longest snake, with specimens regularly measuring over 6m (20ft). It hunts by waiting motionless for a victim to come close, before enveloping it in its coils and squeezing it to death. Prey is then swallowed whole.

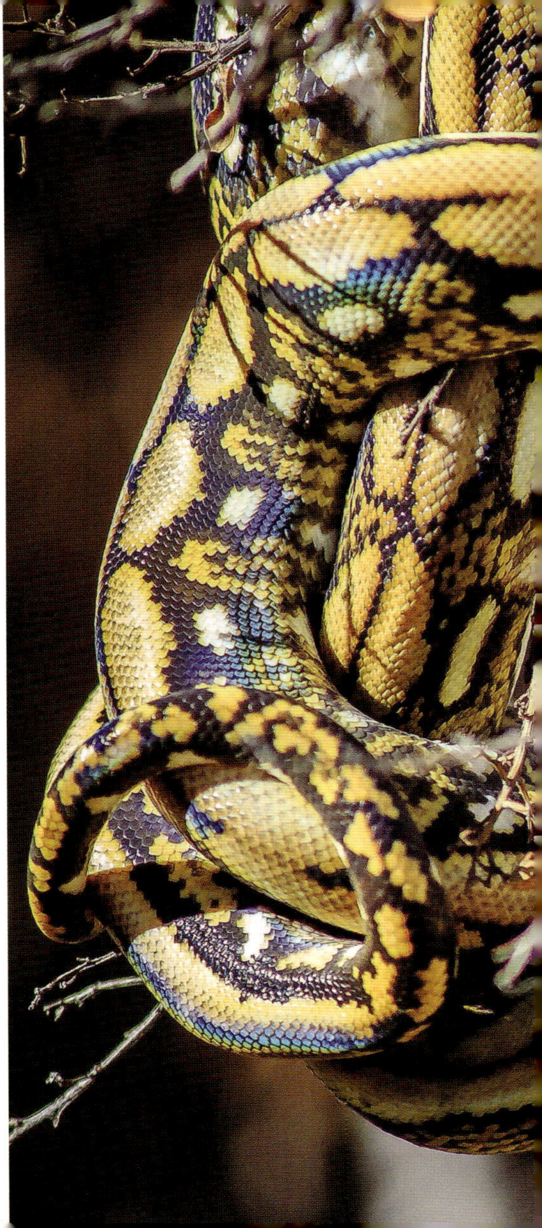

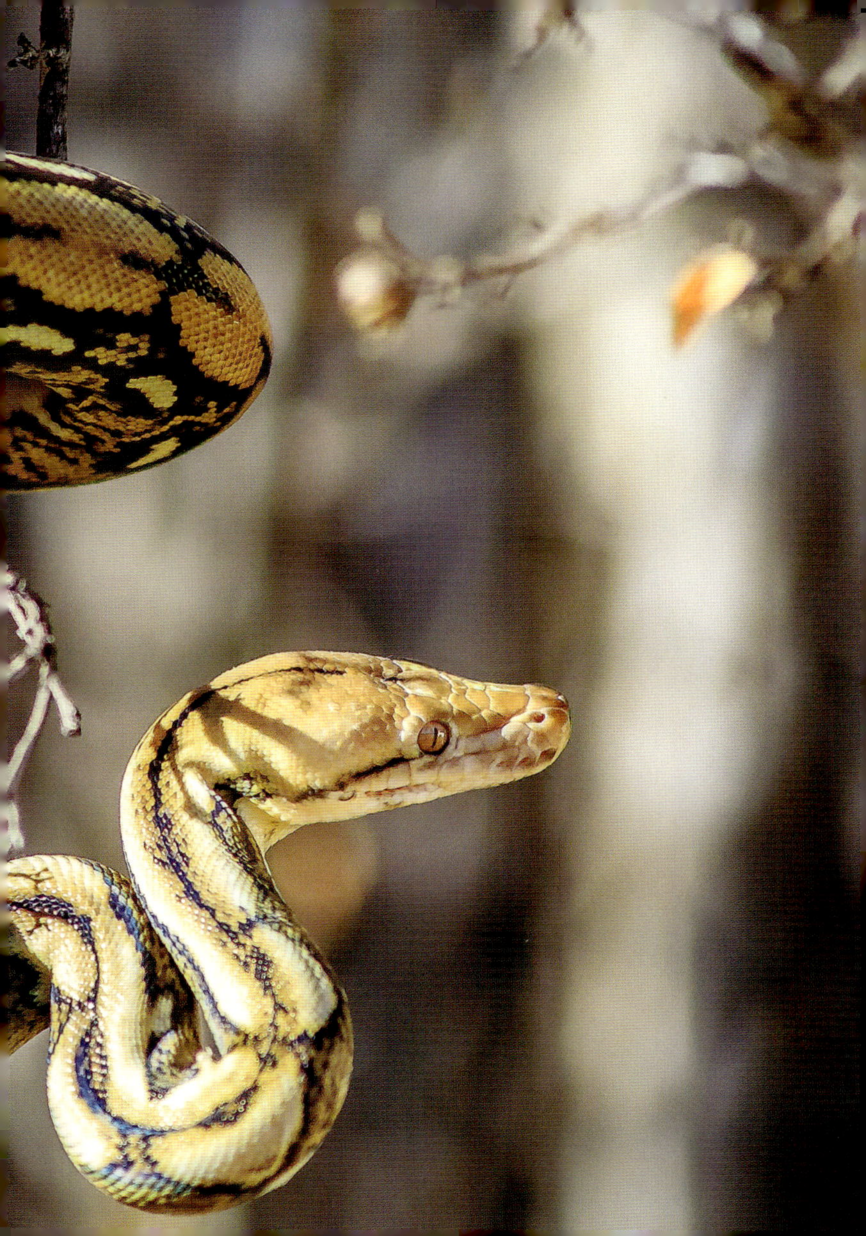

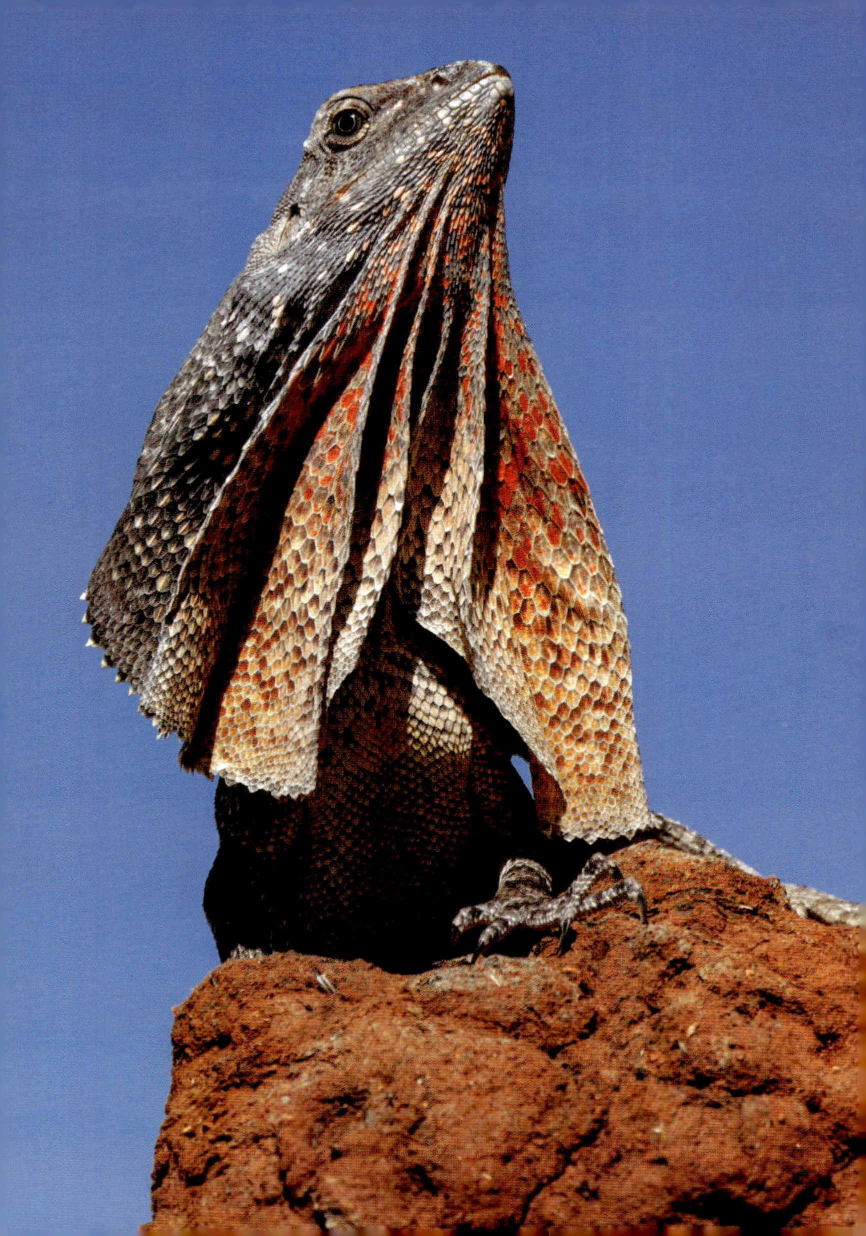

Australasia

Australasia is a continent that encompasses the land masses of Australia, New Zealand, Indonesia and a series of island groups scattered throughout the South Pacific Ocean. Within its area, it has a range of magnificent natural features: tropical rainforest, parched desert, snow-capped mountain ranges and bounteous oceans full of marine creatures of every conceivable shape, size and colour.

Due to its geographical isolation, a wide range of special animal groups have evolved, primarily the marsupials and a number of unique bird families, such as cassowaries and cockatoos. In addition, the platypus and echidna are the only two egg-laying mammals in the world. Australia, in particular, is home to some of the most venomous snakes, spiders and jellyfish inhabiting the planet. You will read about some of them later in this chapter.

New Zealand plays host to a range of unusual birds, such as the kiwi and kakapo, both of which are nocturnal and flightless. In New Guinea, you can find the beautiful birds-of-paradise, and on southern islands albatross and penguins.

Since colonisation by Europeans, many non-native animals have been introduced. This has had a devastating effect of indigenous wildlife, leading to some extinctions and many severe and continued reductions in populations. Fortunately, serious efforts are being made by conservation bodies to conserve what remains, thus giving hope that the region's fascinating wild creatures can consolidate and thrive.

OPPOSITE:

Frilled lizard

So-called because of the large frill around its neck, these lizards are endemic to northern Australia and southern New Guinea, and can reach lengths of 85cm (2.8ft).

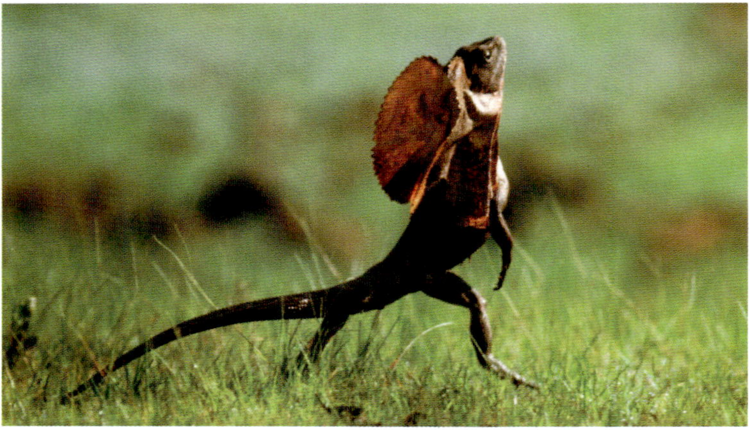

ALL PHOTOGRAPHS:

Frilled lizard

The bright red frill around the neck is supported by a ring of cartilage spines which can be raised when the lizard feels threatened, puts on a territorial display or during courtship. Although tree dwelling, frilled lizards have an ability to move across ground on their hind legs at speeds of up to 25km/h (15.5mph).

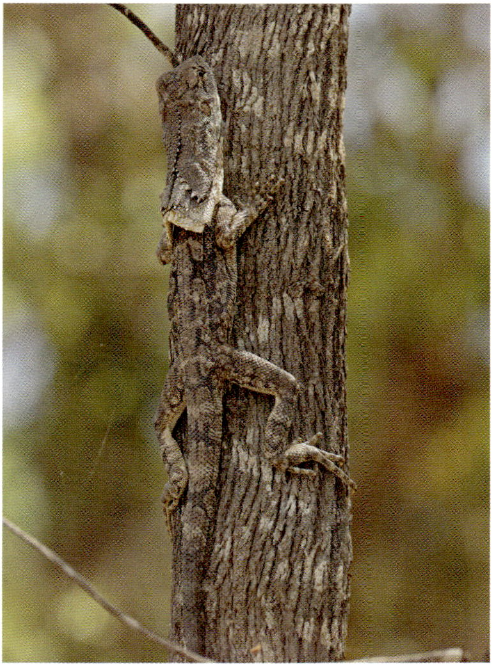

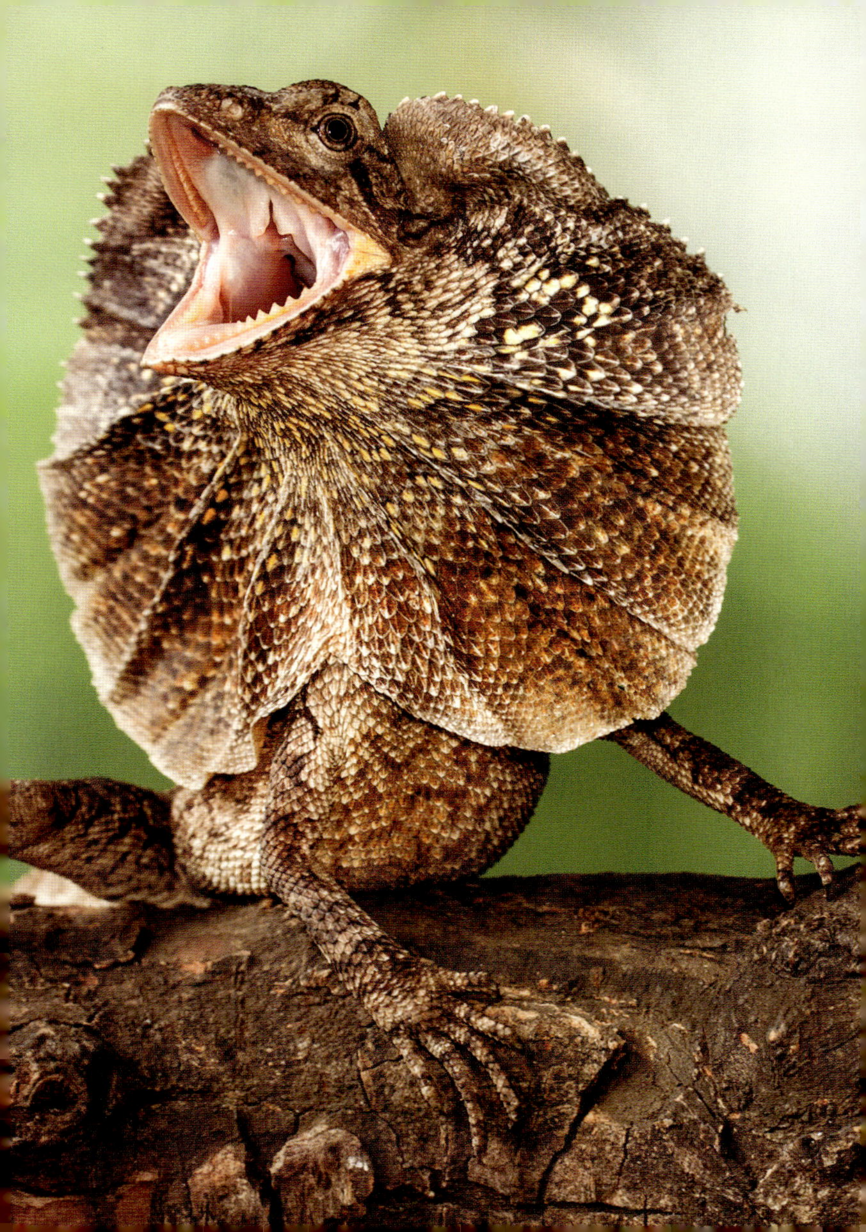

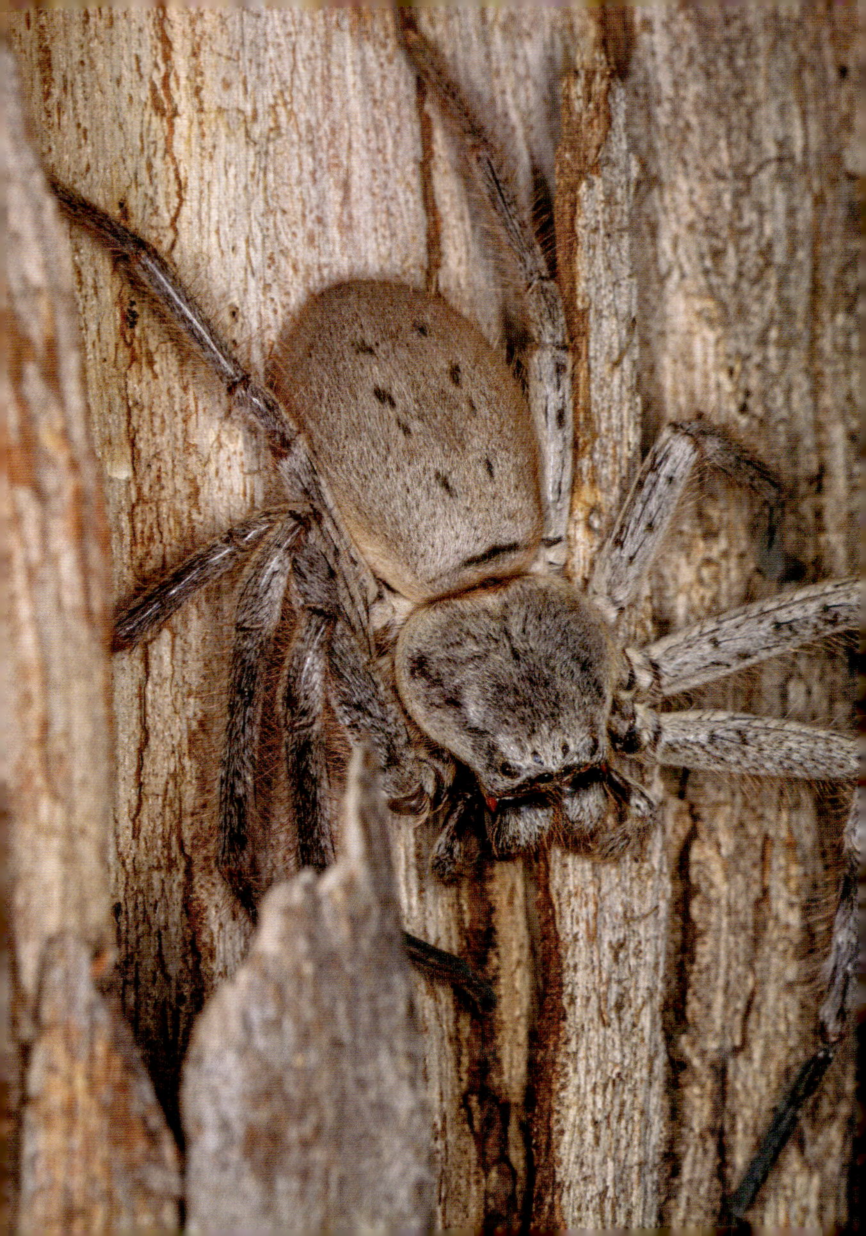

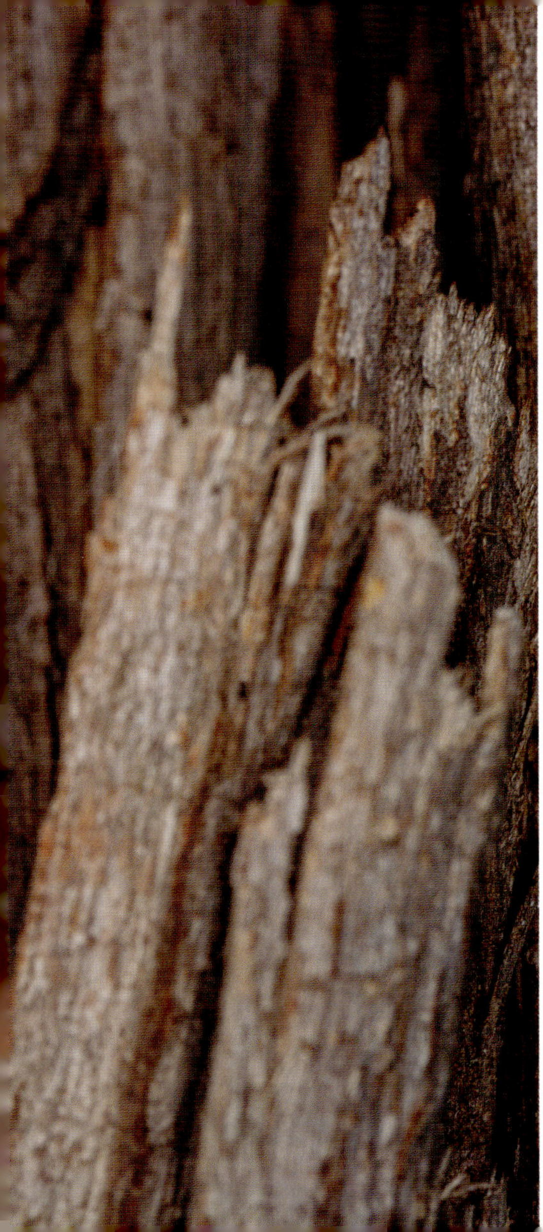

Huntsman spider

There are several species of huntsman spiders widely distributed across Australia. They have a body length of 2cm (0.8in), but their leg span can be up to 15cm (6in). They do not build webs, preferring to actively hunting their invertebrate prey. They pose little threat to humans despite quite regularly coming into houses.

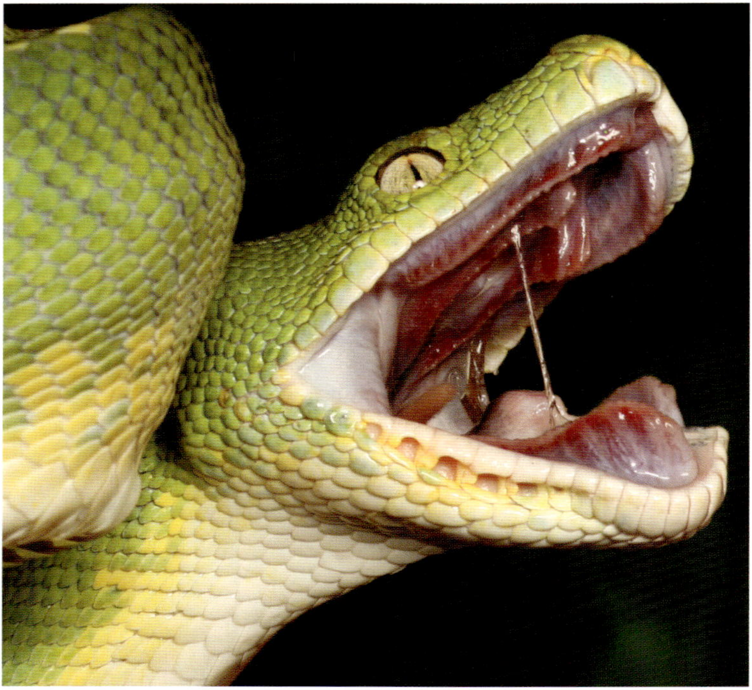

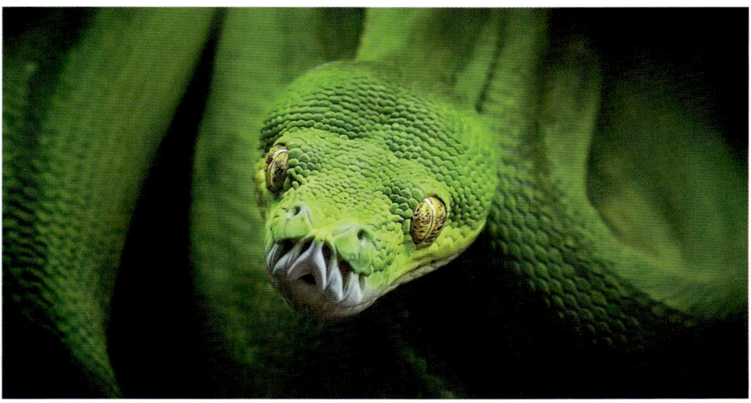

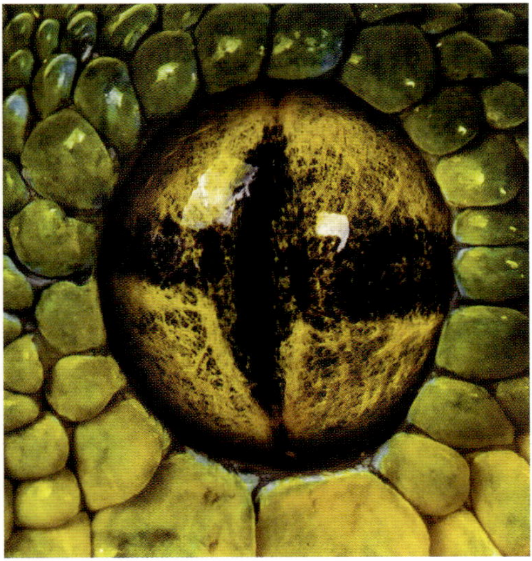

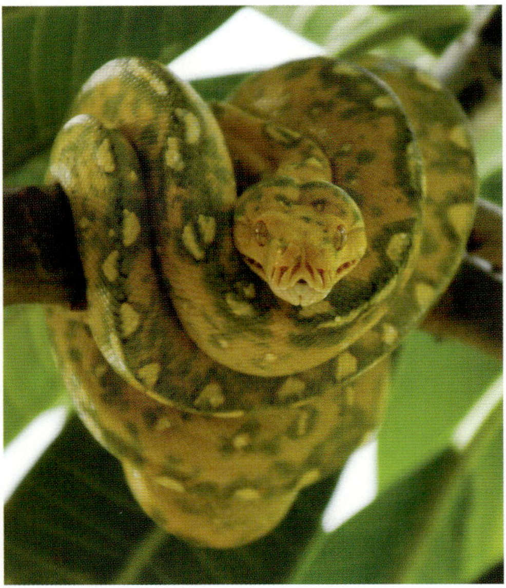

ALL PHOTOGRAPHS:

Green tree python

This non-venomous, tree-dwelling snake can be found in the rainforests of northern Australia, New Guinea and some nearby island groups. It is generally bright green in colour, which provides excellent camouflage when resting in trees. It reaches an average length of 1.5m (5ft), and can live for 20 years. Although sometimes an active hunter, these snakes spend a lot of time coiled around tree branches. There, they catch food by sitting motionless and wiggling the tip of their tail, luring unsuspecting prey to within striking distance. Their main diet comprises lizards, birds and small mammals.

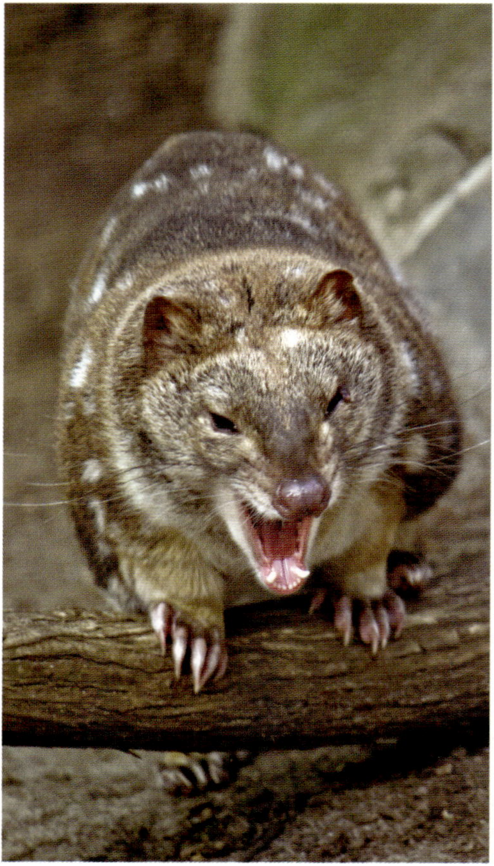

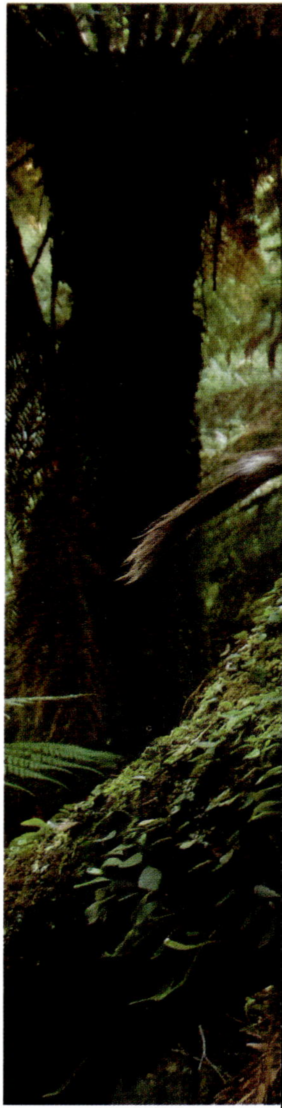

ALL PHOTOGRAPHS:

Quoll

Six species exist, all living in distinct geographic zones of Australia and New Guinea. Quolls are carnivorous nocturnal marsupials with a wide-ranging diet of small mammals, birds, frogs and invertebrates. As a marsupial, females develop a pouch when breeding, which can accommodate six young.

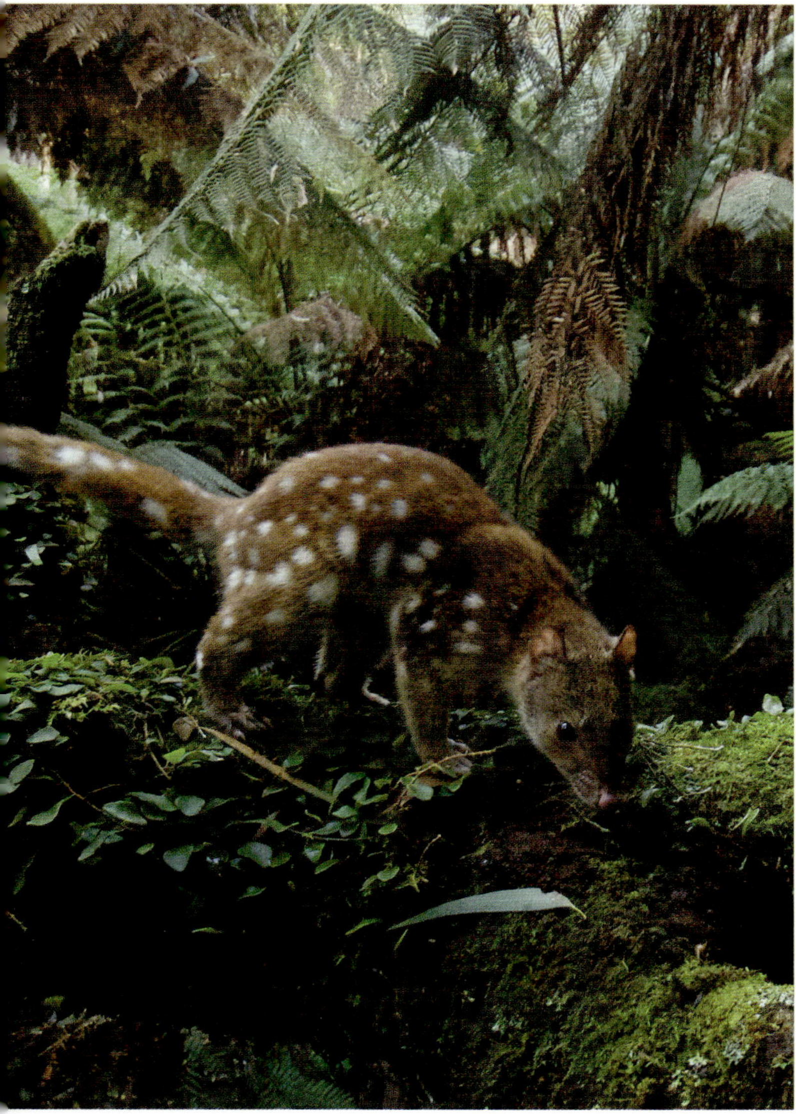

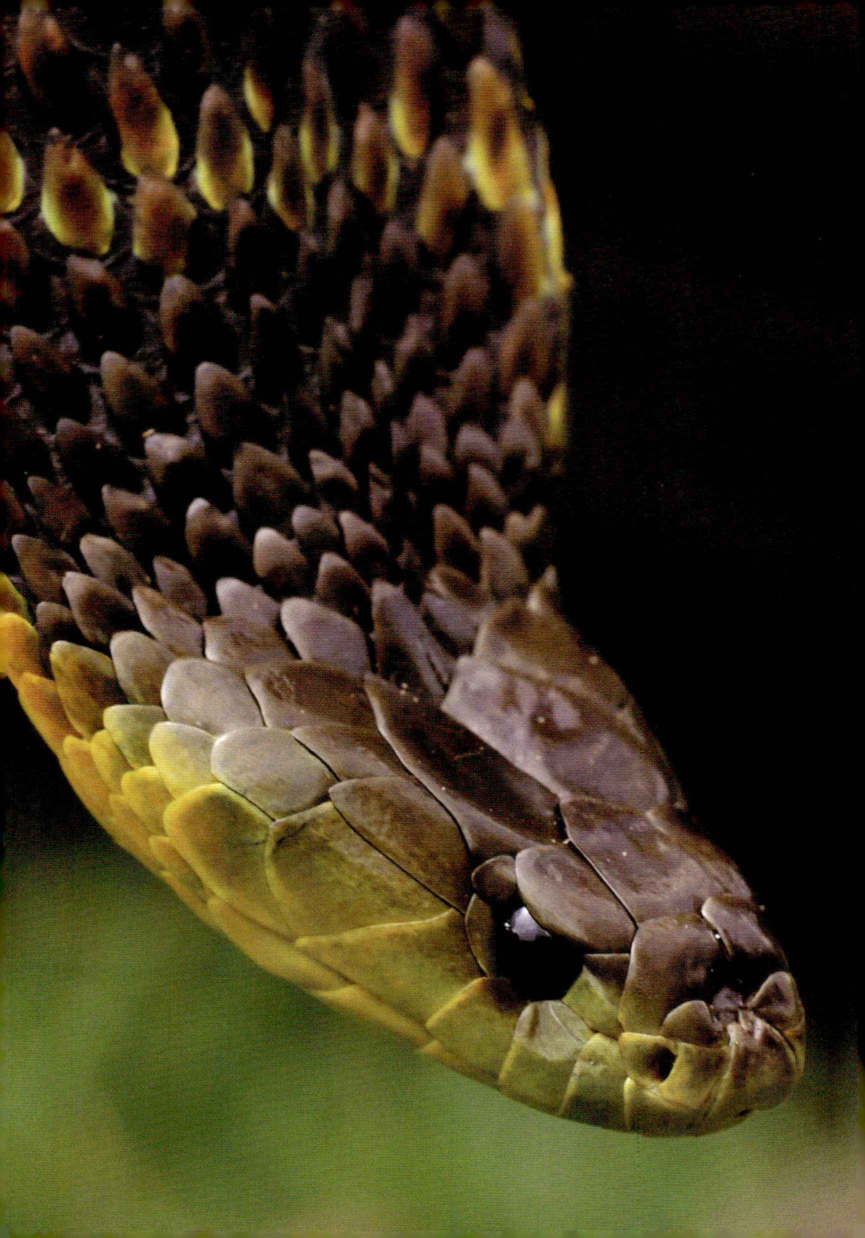

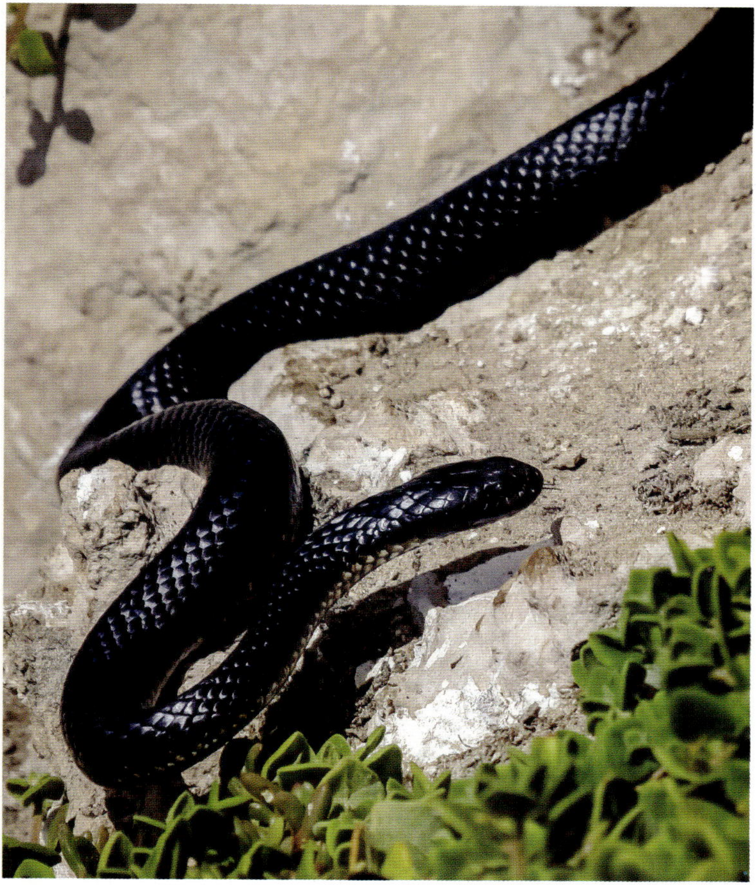

ALL PHOTOGRAPHS:

Tiger snake

A highly venomous Australian snake with a number of distinctive geographic forms. Some forms are clearly banded, thus giving the species its name, others are completely black. It is responsible for 17 per cent of recorded snake bites, but these are rarely fatal thanks to effective and widely available antivenom.

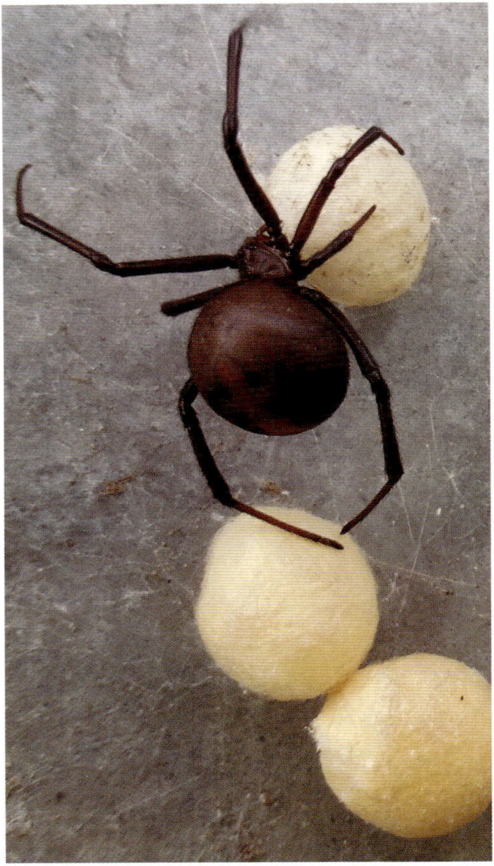

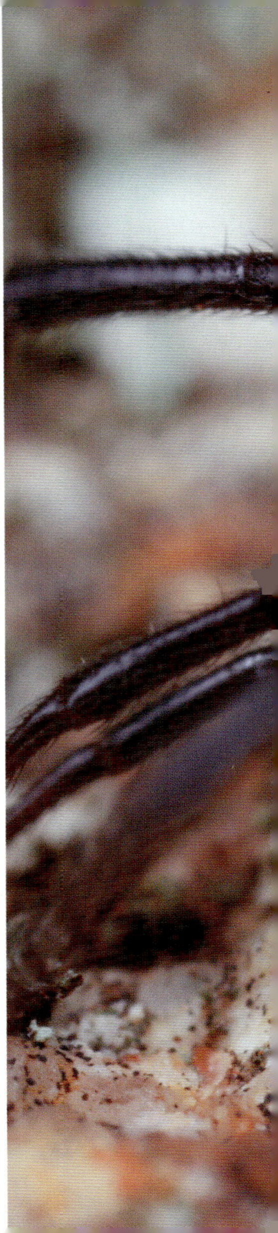

ALL PHOTOGRAPHS:

Redback spider

A widespread Australian spider that is considered one of the most dangerous of them all. The female, which is about 10mm (0.4in) long, has a round black body with a distinctive red marking on the back and underside. Males are much smaller, approximately 4mm (0.16in).

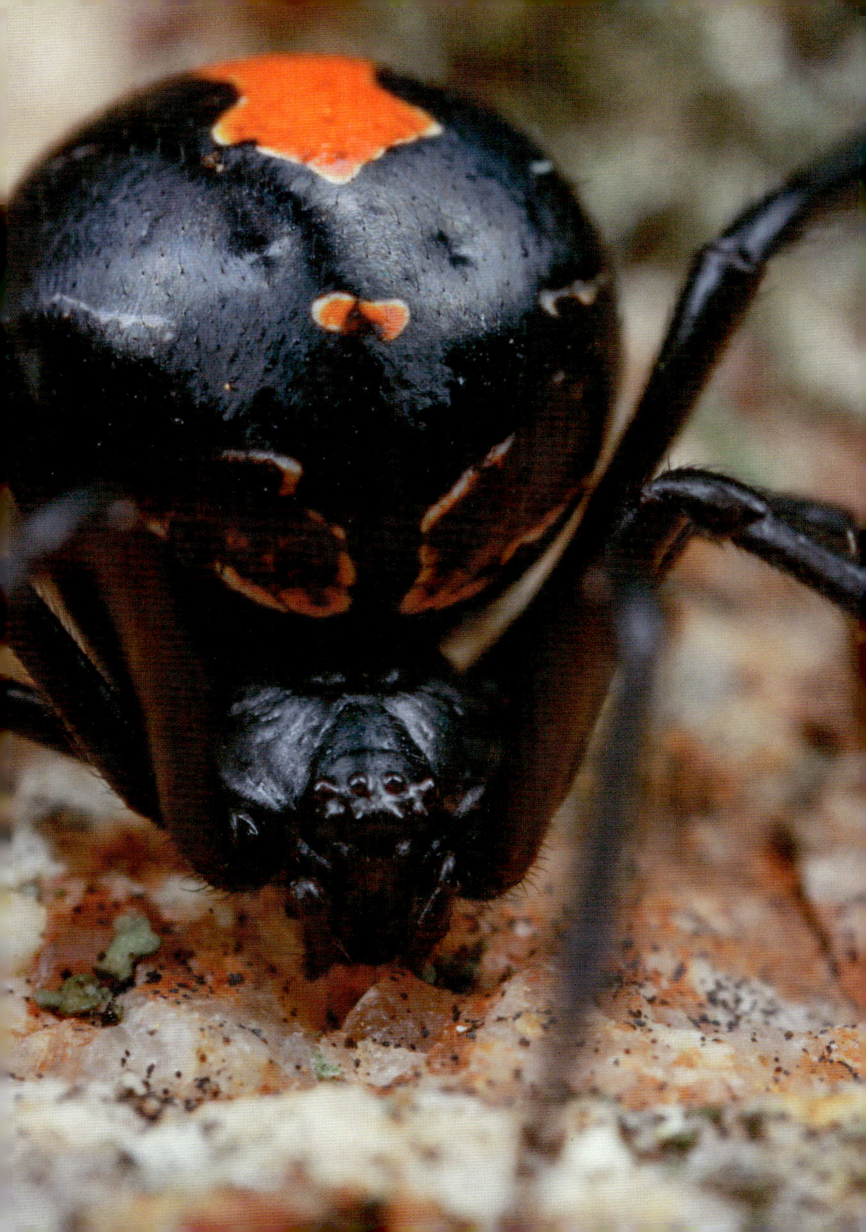

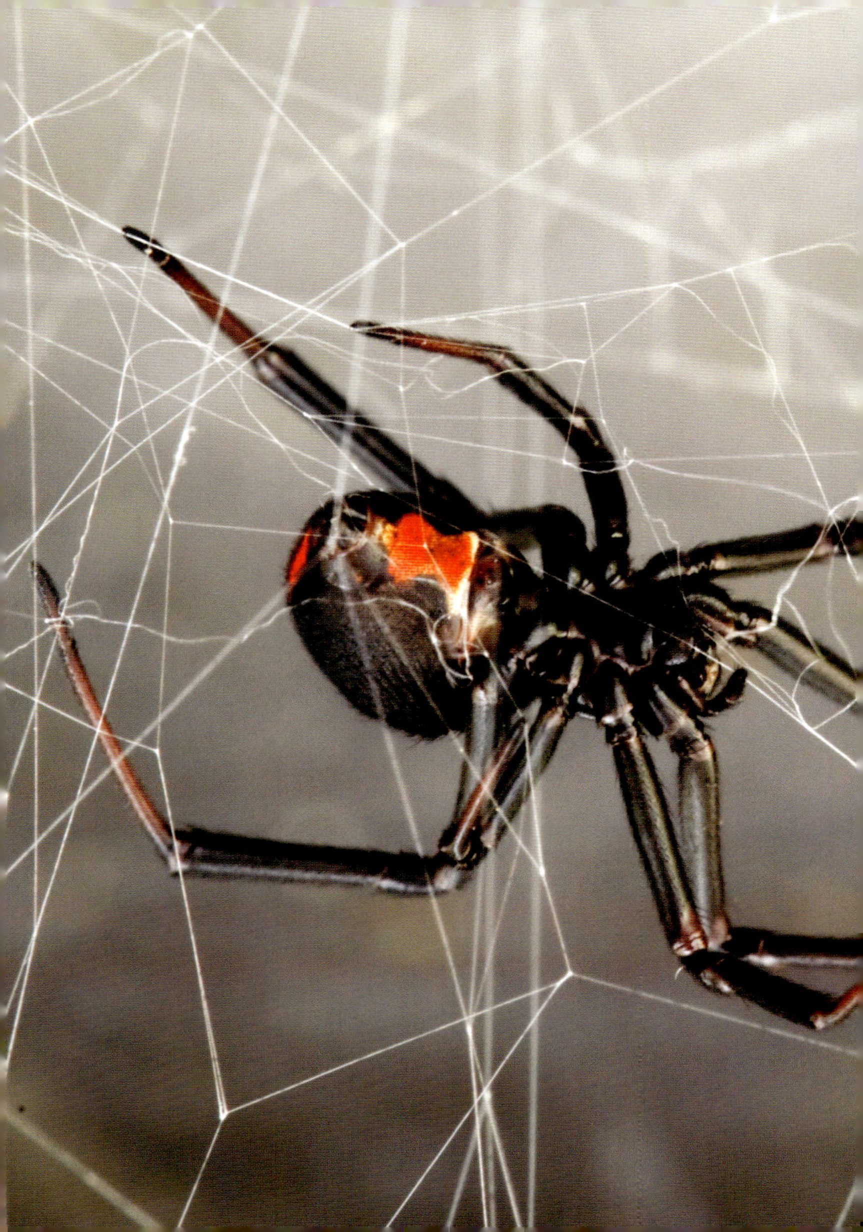

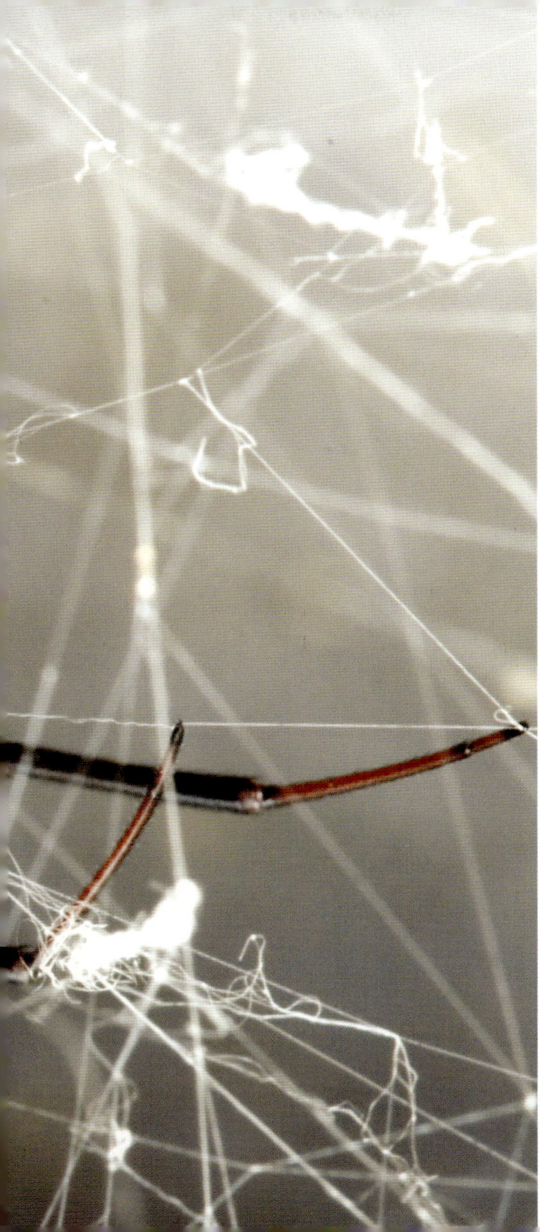

Redback spider in a web
Redbacks prefer to make their webs in warm locations, which often includes human habitations. The female is credited with biting several thousand people every year. The bite causes severe pain which can last for days but is rarely fatal, although a small number require antivenom.

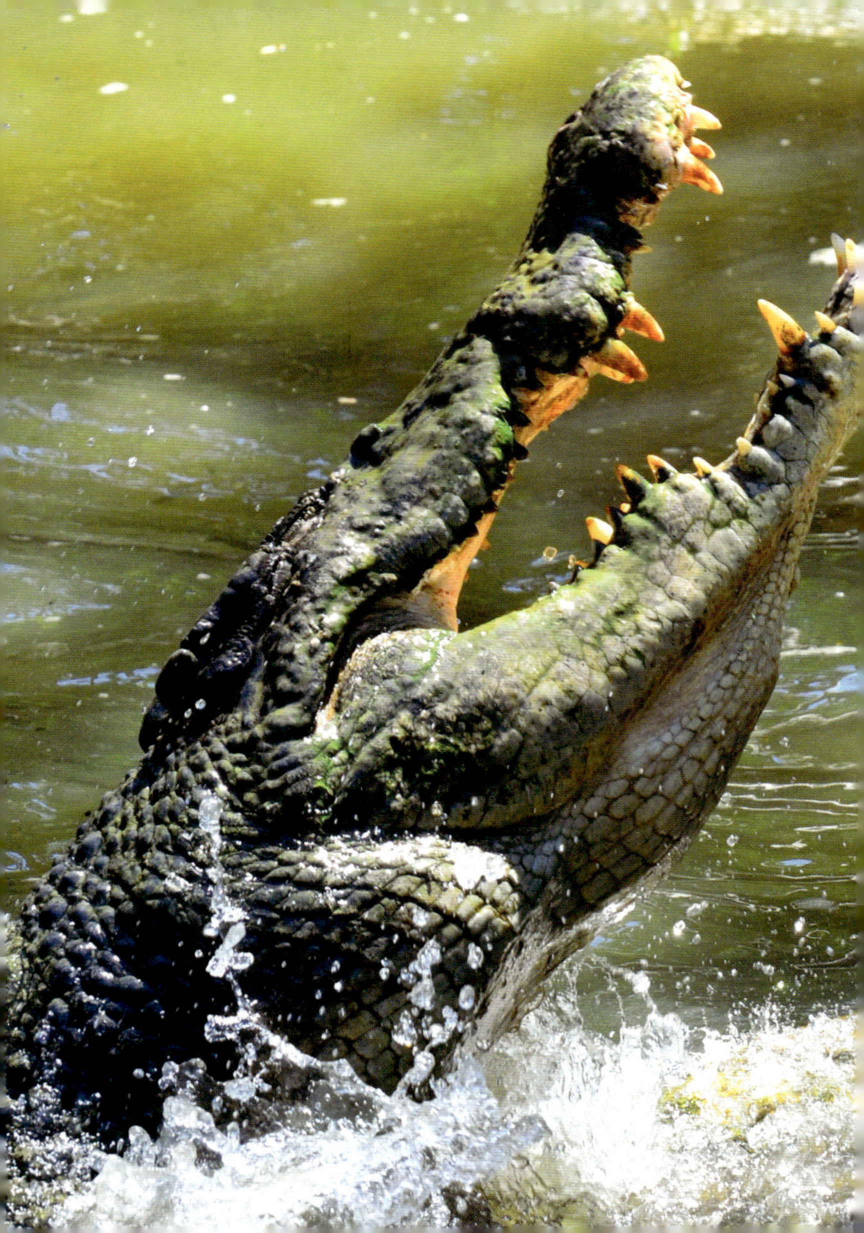

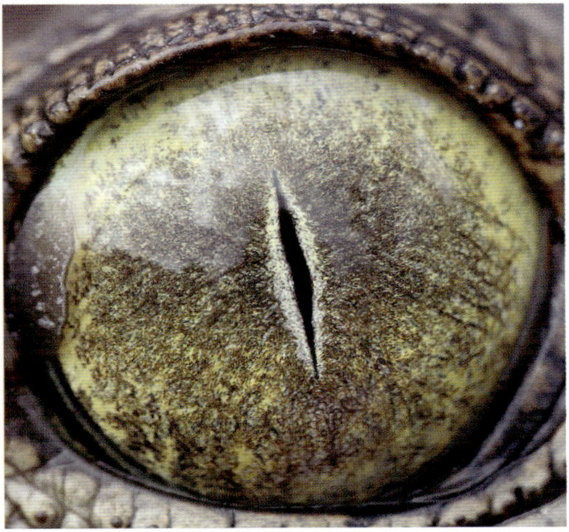

ALL PHOTOGRAPHS:

Saltwater crocodile

A thriving species in the northern part of the region, it inhabits brackish wetlands and river systems, but can also swim long distances in saltwater. Males reach lengths of 6m (19.6ft) with females being about half that size. The Australian population is estimated to be 100,000–200,000 adults.

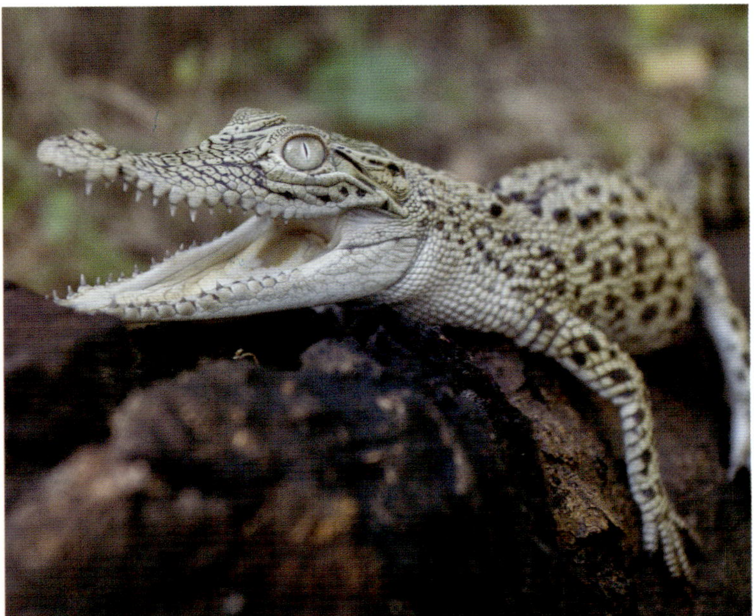

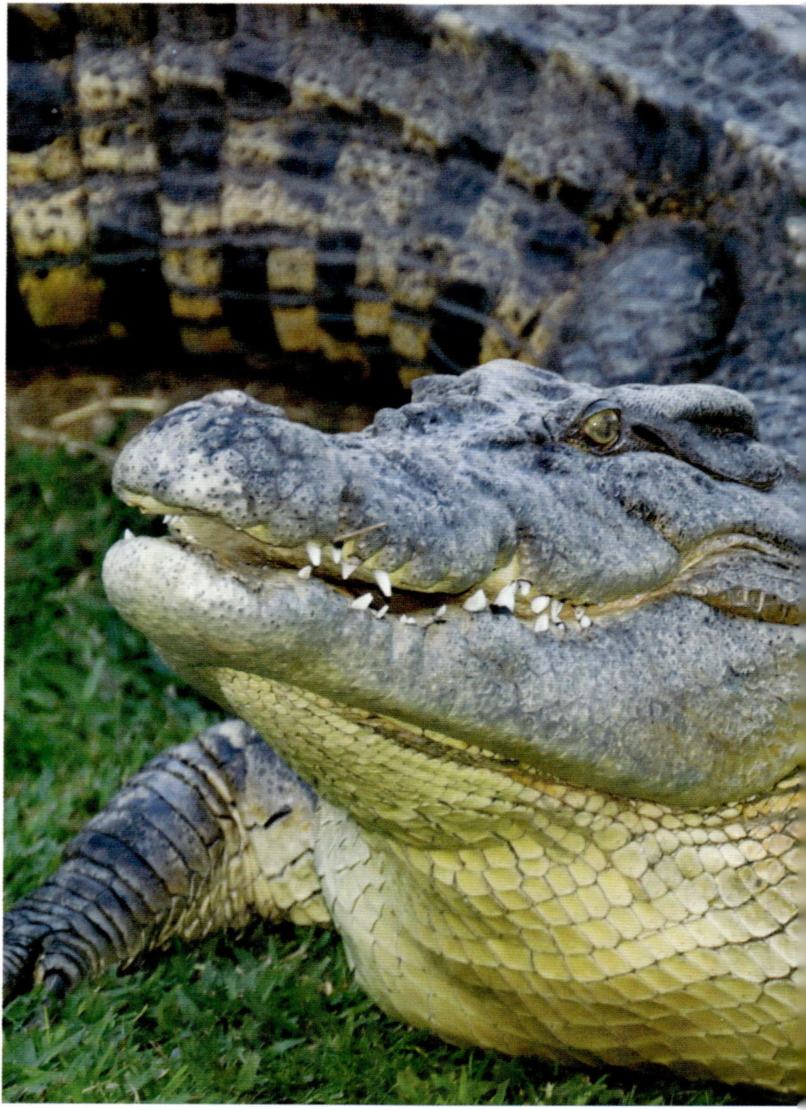

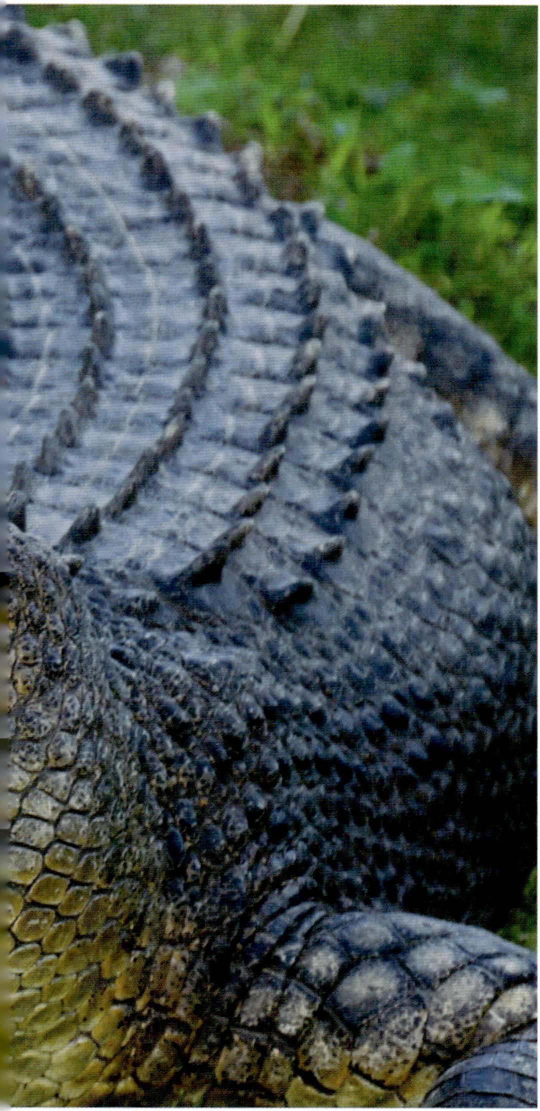

Saltwater crocodile

These marvellously well-adapted crocodiles have the most powerful bite of any living animal. They have a wide-ranging diet that includes fish, crabs, birds and small mammals. Territorial males have shown a tendency to attack boats and even humans, although fatal attacks are rare with only one or two reported every year.

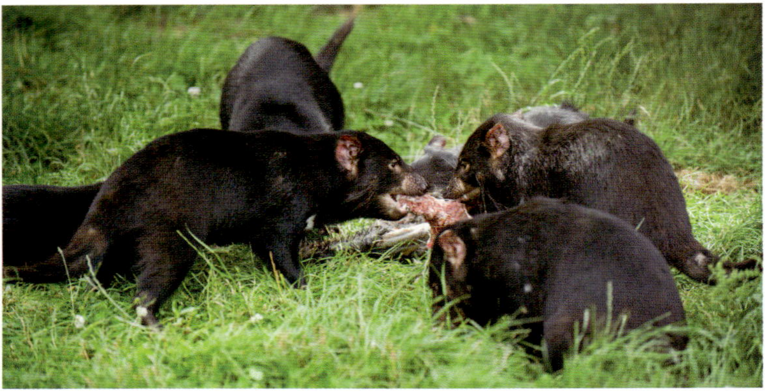

ALL PHOTOGRAPHS:

Tasmanian devil

The Tasmanian devil is widely distributed on the island of Tasmania and has recently been reintroduced to the Australian mainland. It is a carnivorous marsupial with an overall length of around 90cm (35in). The fur is black with a white patch on the chest.

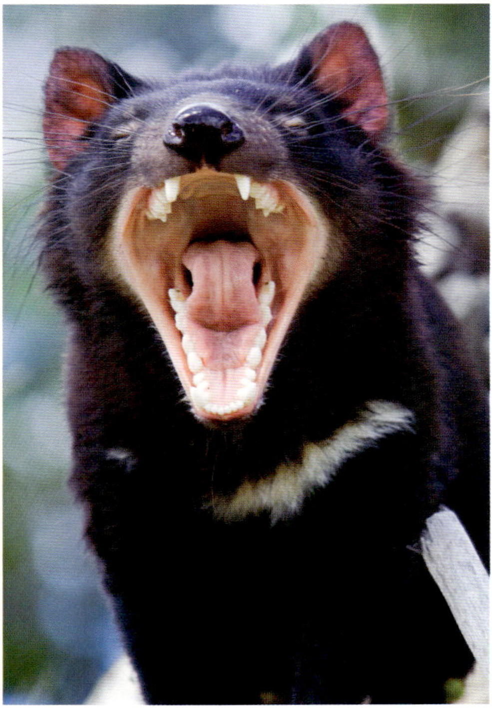

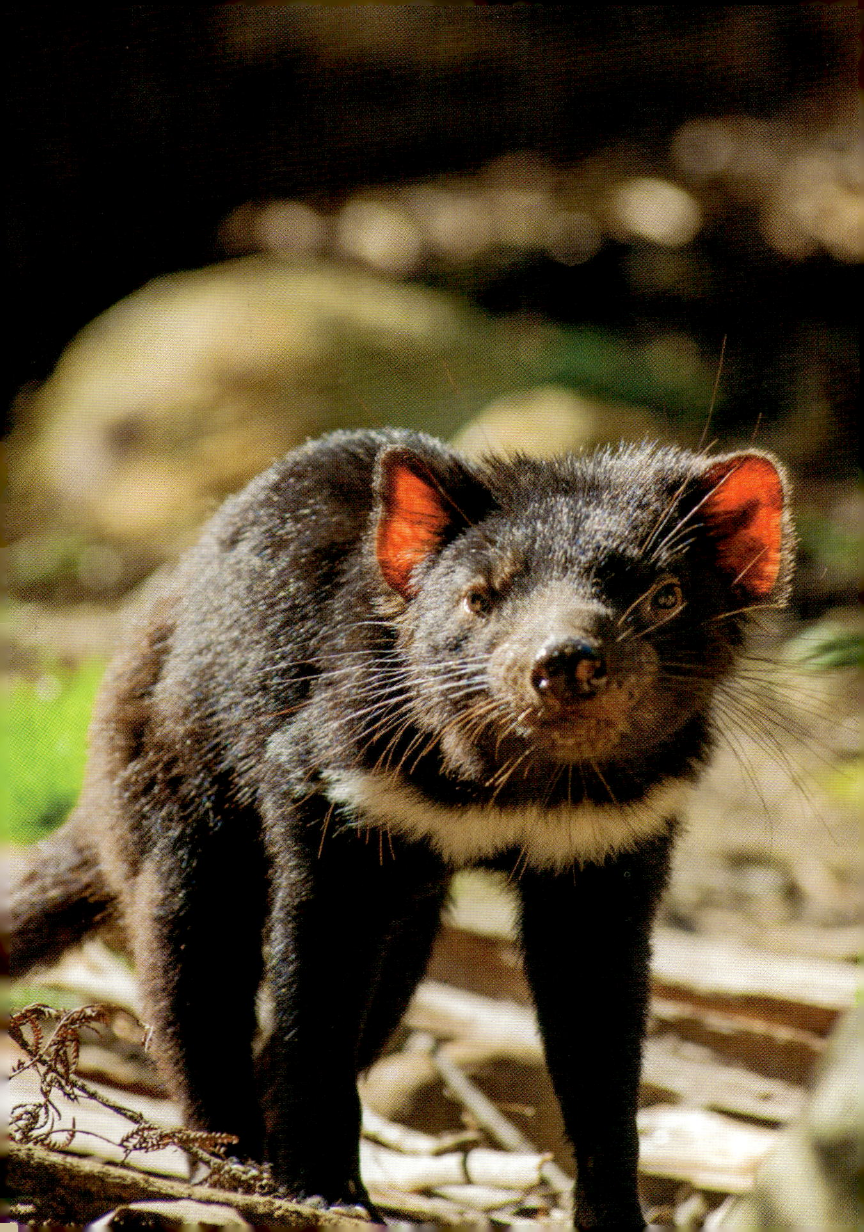

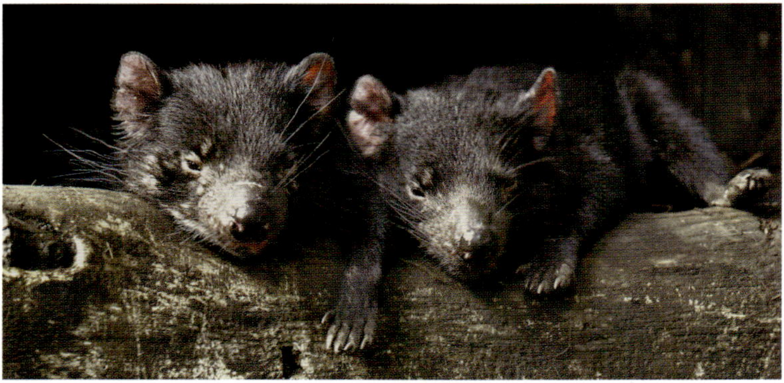

ABOVE AND RIGHT:

Tasmanian devil

Tasmanian devils are nocturnal hunters, taking prey such as wombats, small wallabies and kangaroos. They also commonly feed on roadkill and carrion. Young animals are nurtured in a pouch where they will stay for 16 weeks after birth. After weaning, they remain in a safe den for a further three months.

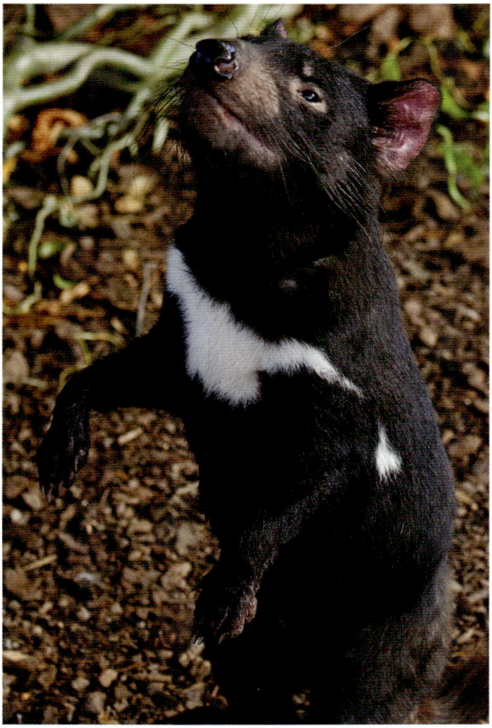

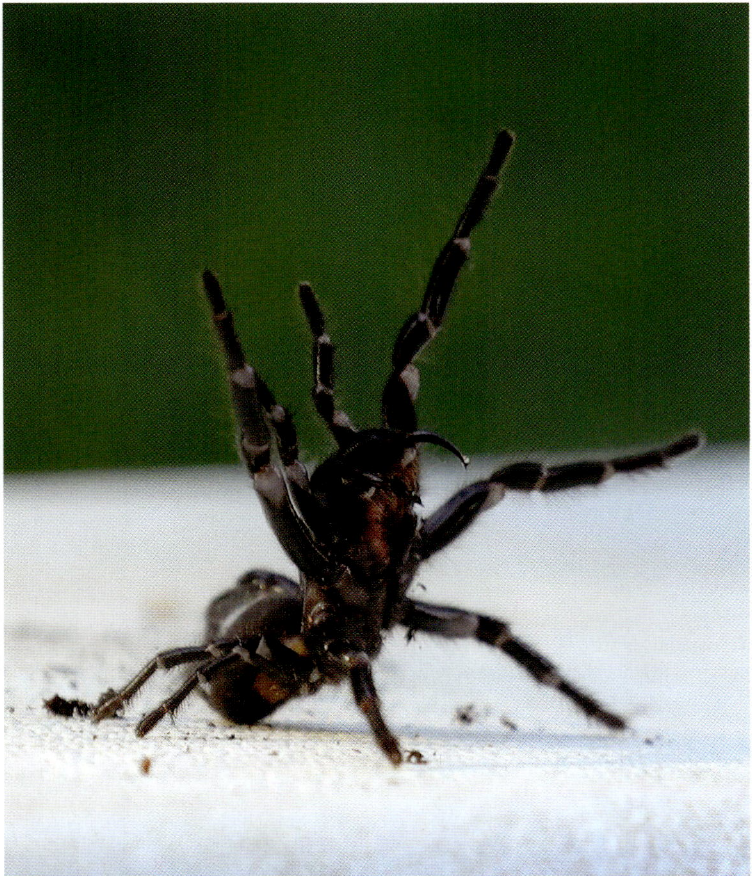

ABOVE AND OVERLEAF:
Sydney funnelweb spider
This particular species, ranging in size from 2–5cm (0.8–2in), makes its home in and around the Sydney area of Australia. It is highly venomous with fangs that can pierce human skin. For this reason, bites are treated as a medical emergency. When threatened, the spider rears up and displays those formidable fangs.

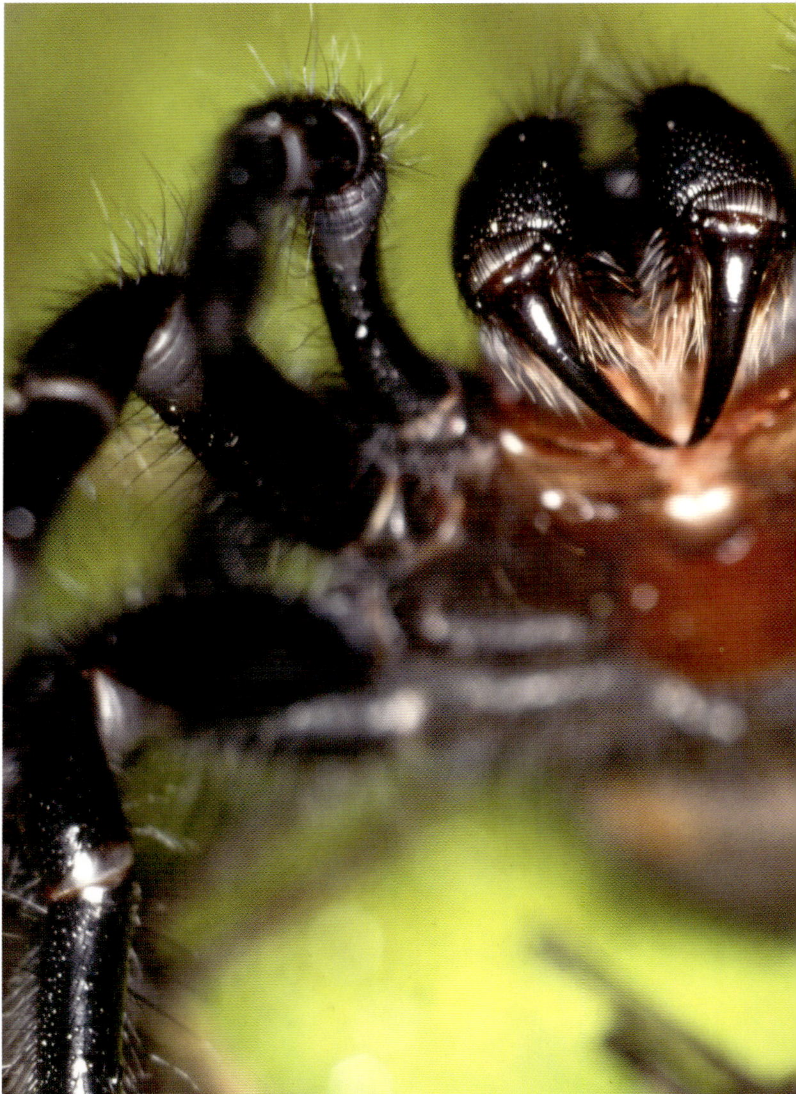

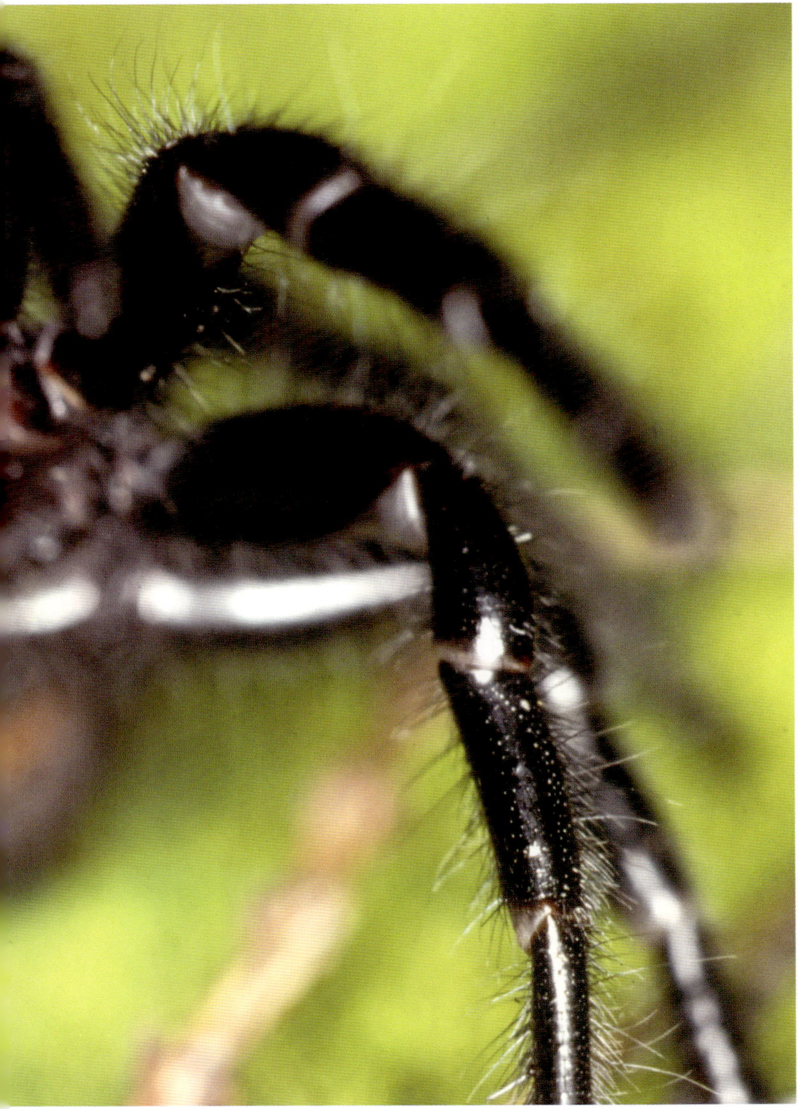

Taipan snake

Three species of taipan are endemic to Australia, averaging nearly 2m (6.5ft) in length. It is a highly venomous snake feeding on small rodents and mammals. Inland taipans, inhabiting areas of central eastern Australia, are shy creatures, whereas the coastal taipans, when threatened, can be a more aggressive species.

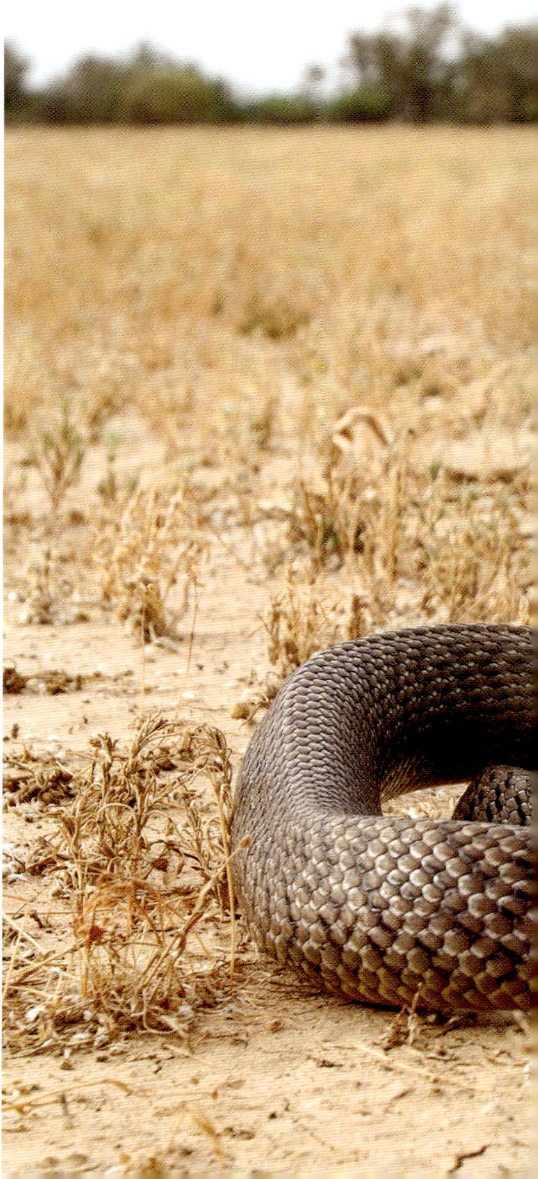

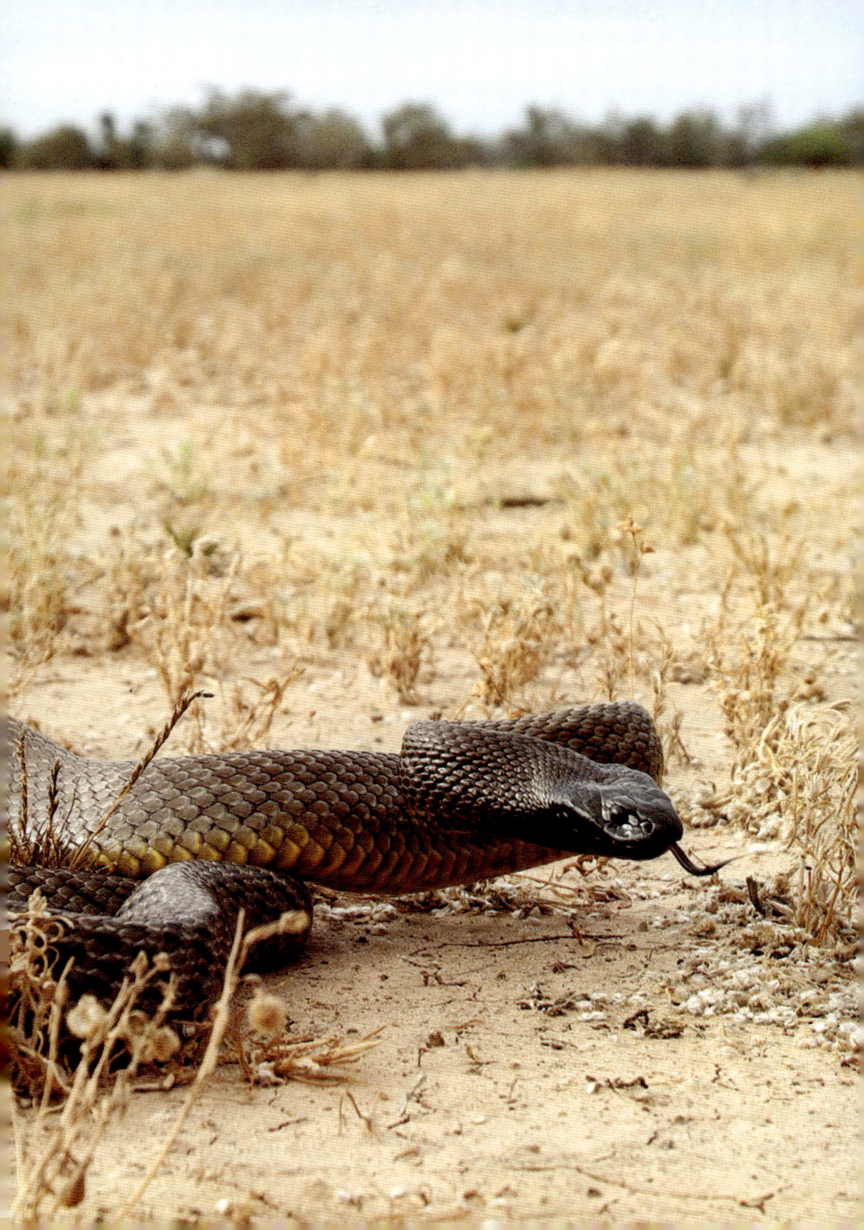

Wolf spider

A widespread group of species, commonly found in human habitation. Some species are quite large, with a leg span of 10cm (4in) and, although venomous, pose little threat to humans. They hunt by actively stalking prey and are considered a beneficial arachnid, ridding houses of less desirable invertebrates.

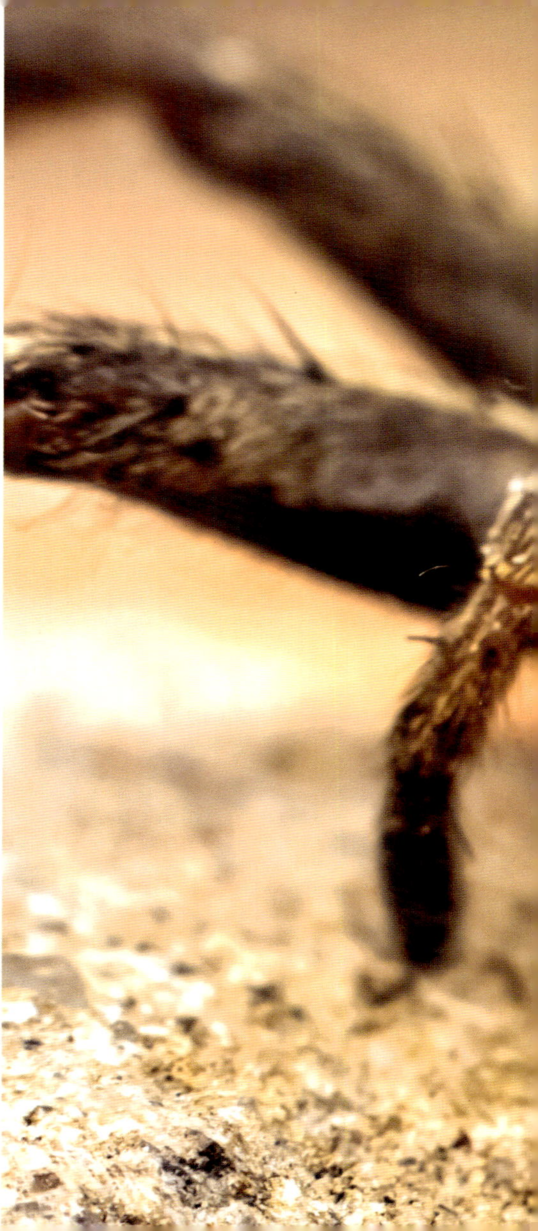

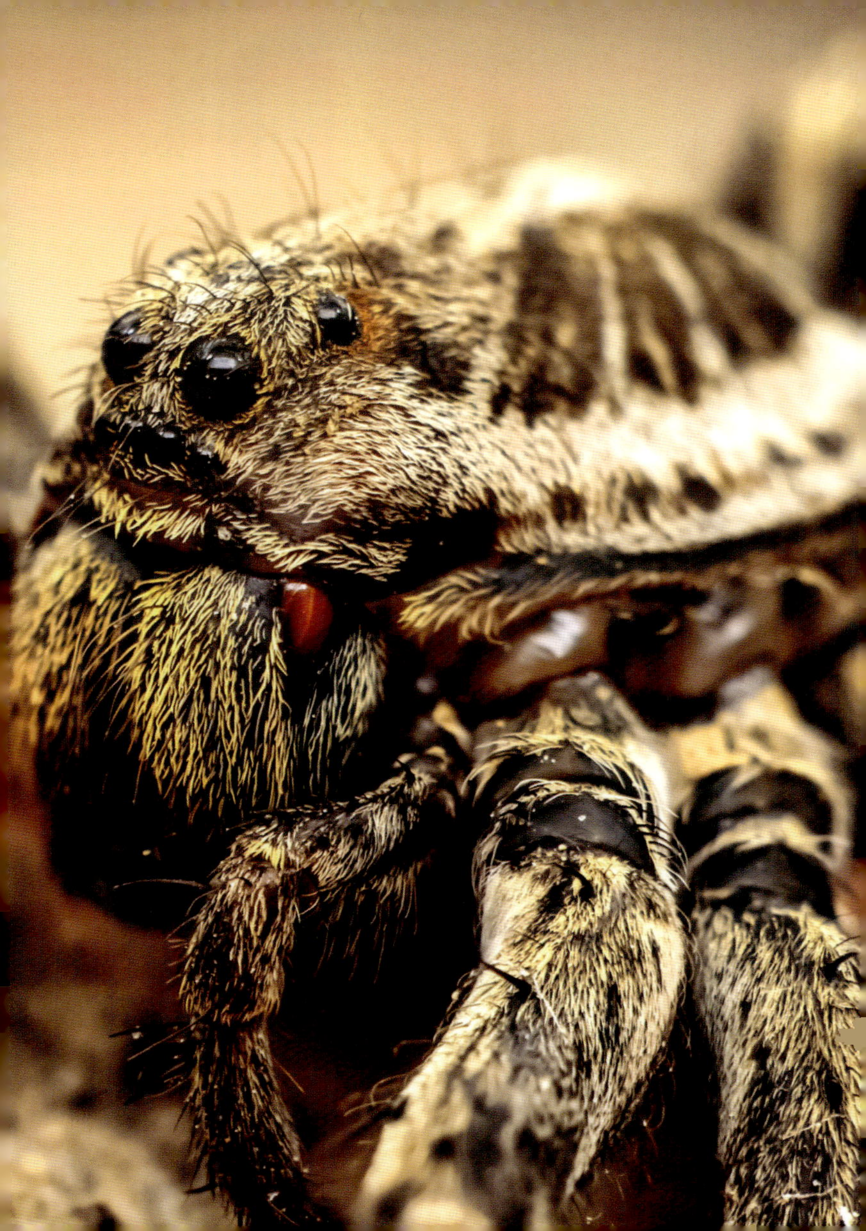

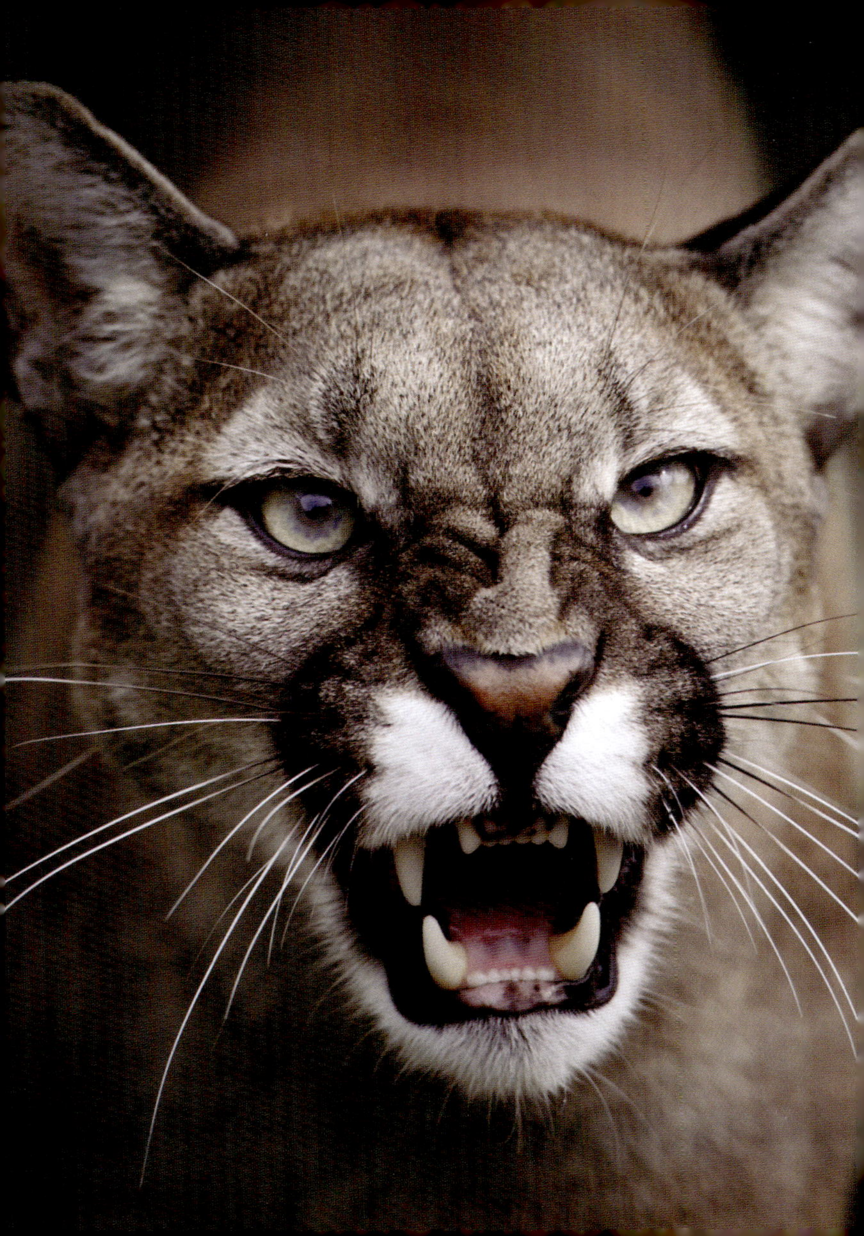

North America

The continent of North America encompasses a total of 23 countries with the largest two being the USA and Canada. It contains the world's largest island, Greenland, and the world's hottest region, Death Valley in California. It has a total population of some 529 million people, many of who live in large cities concentrated towards coastal regions. This leaves much of the interior as true wilderness, often protected as national parks – vital habitats for a wide range of indigenous wildlife.

North America's size and geographic position results in a wide variety of climatic zones, from Arctic tundra, and both broadleaf and coniferous forest, to desert, prairie grasslands and tropical rainforest. Among its myriad wildlife spectacles are the winter roosting of millions of monarch butterflies in the high forests of Mexico, the migratory movement of birds of prey along the eastern USA coastlands, and seasonal movements of humpback whales along Pacific coasts.

The overall list of wild creatures inhabiting the continent is impressive: over 450 mammal species, including pumas and bears; over 900 species of birds, including a range of hummingbirds, wild turkeys, Californian condors and the emblematic bald eagle; close to 1,000 species of reptiles and amphibians, which includes the American alligator, various species of rattlesnake, bullfrogs and salamanders, and 400 species of spider and scorpion. A taste of this naturalist's treasure trove is presented on the following pages.

OPPOSITE:

Puma

The puma, often called the cougar or mountain lion, ranges widely over the Americas. It is the world's fourth-largest cat species, with adult males being 2.5m (8ft) long.

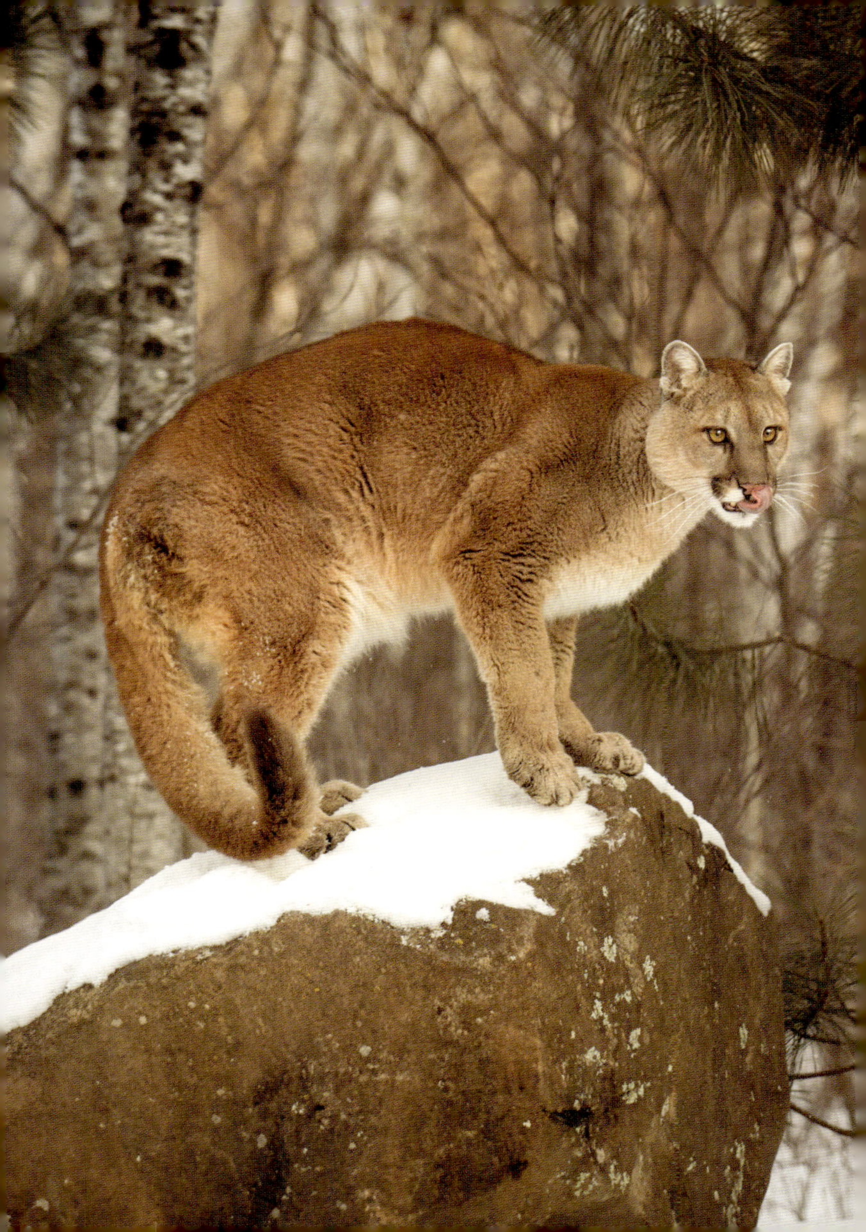

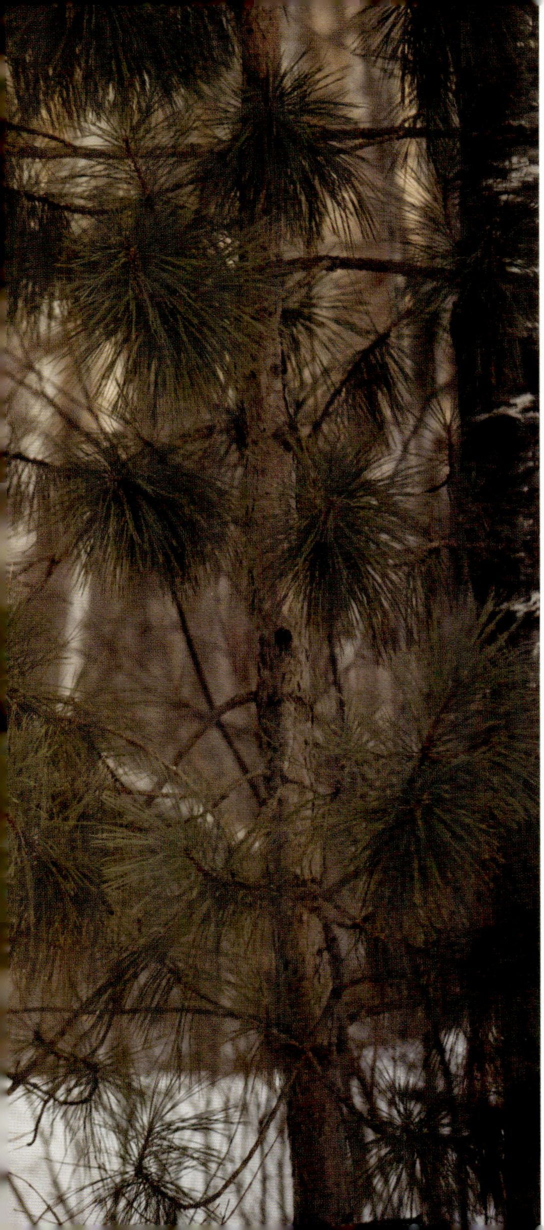

Puma

Despite numerous encounters between humans and this majestic cat, attacks are very rare. Moreover, the continued decline of the species in North America can be directly attributed to hunting, which is still legal in every US state except California, and other aspects of human encroachment.

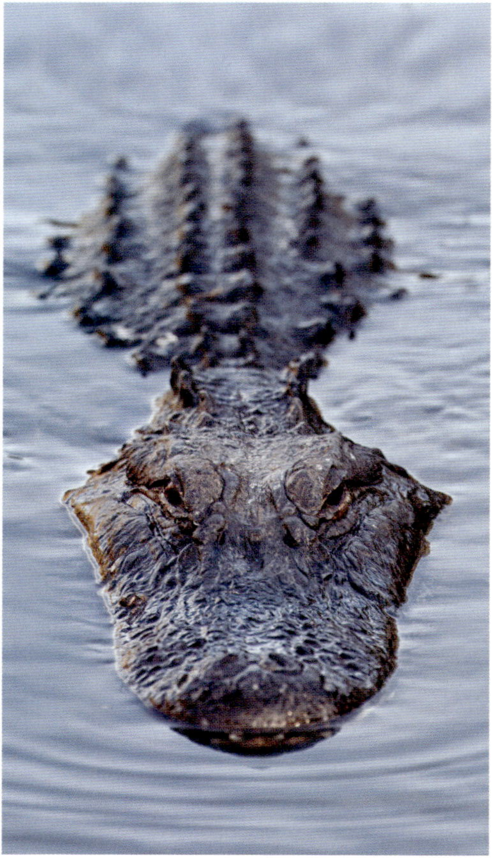

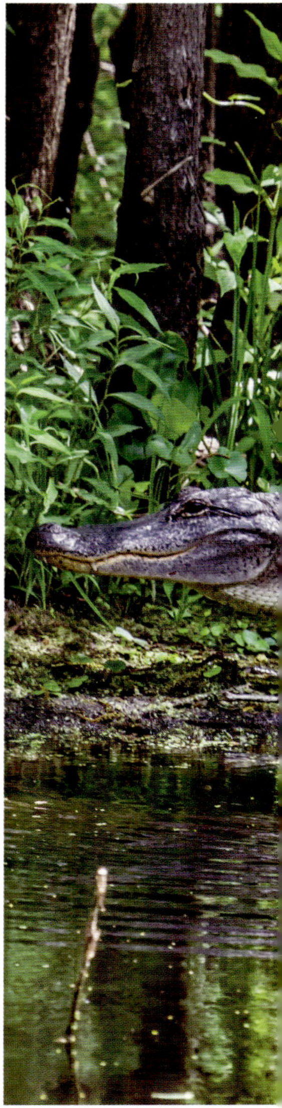

ALL PHOTOGRAPHS AND OVERLEAF:

American alligator

The American alligator can be found in southern states of the USA where they inhabit freshwater swamplands, lakes and marshes. It is usually a solitary animal, but is commercially farmed for its hide, when large numbers will gather around feeding stations as pictured in the image overleaf.

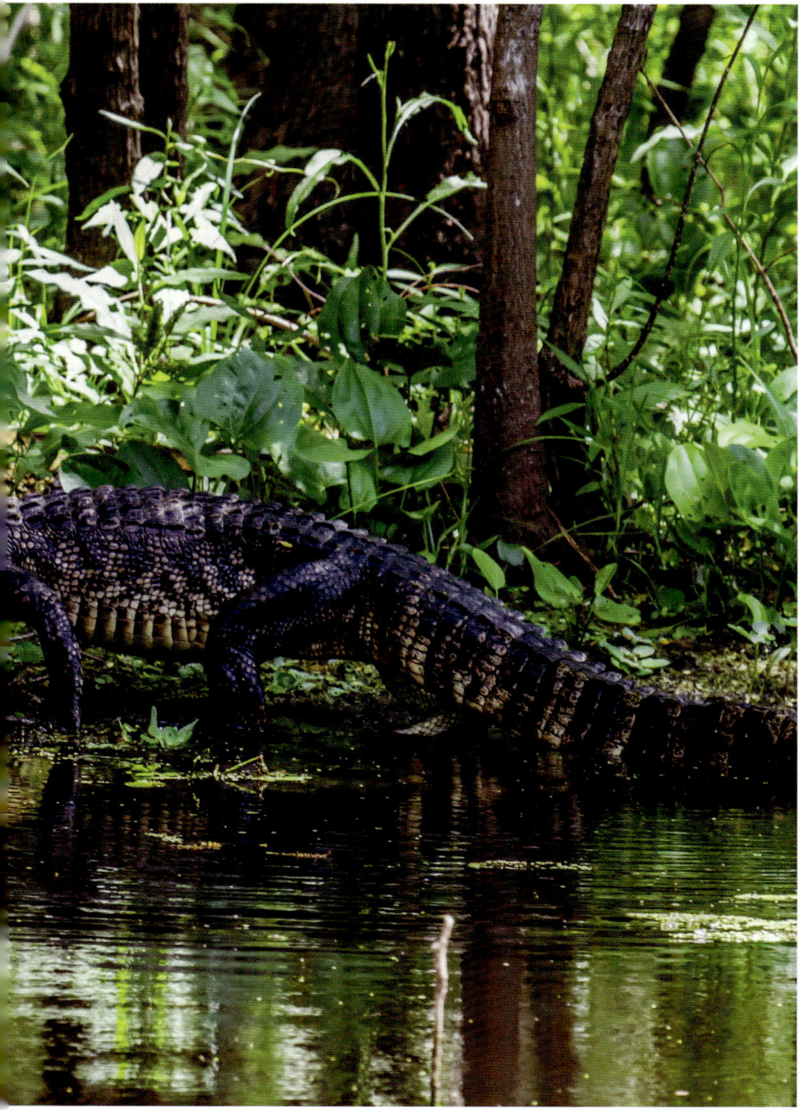

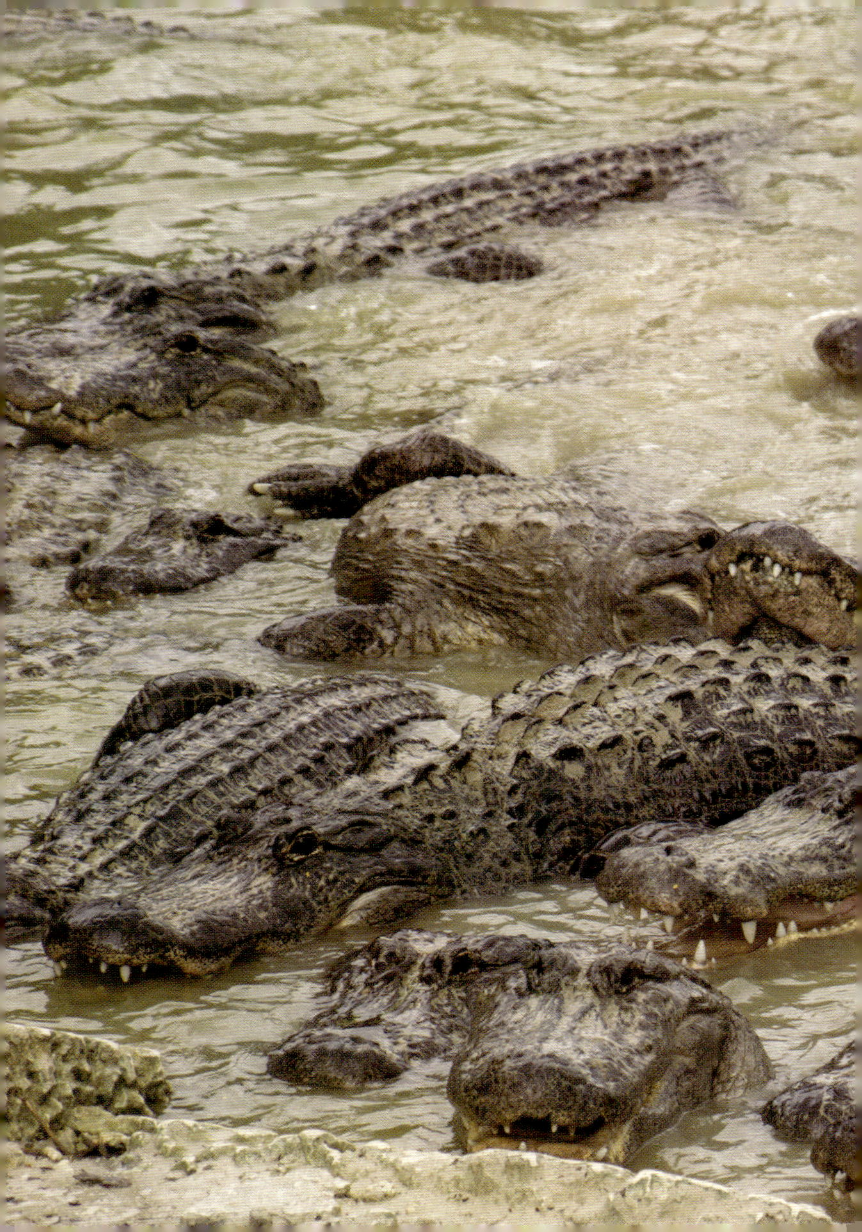

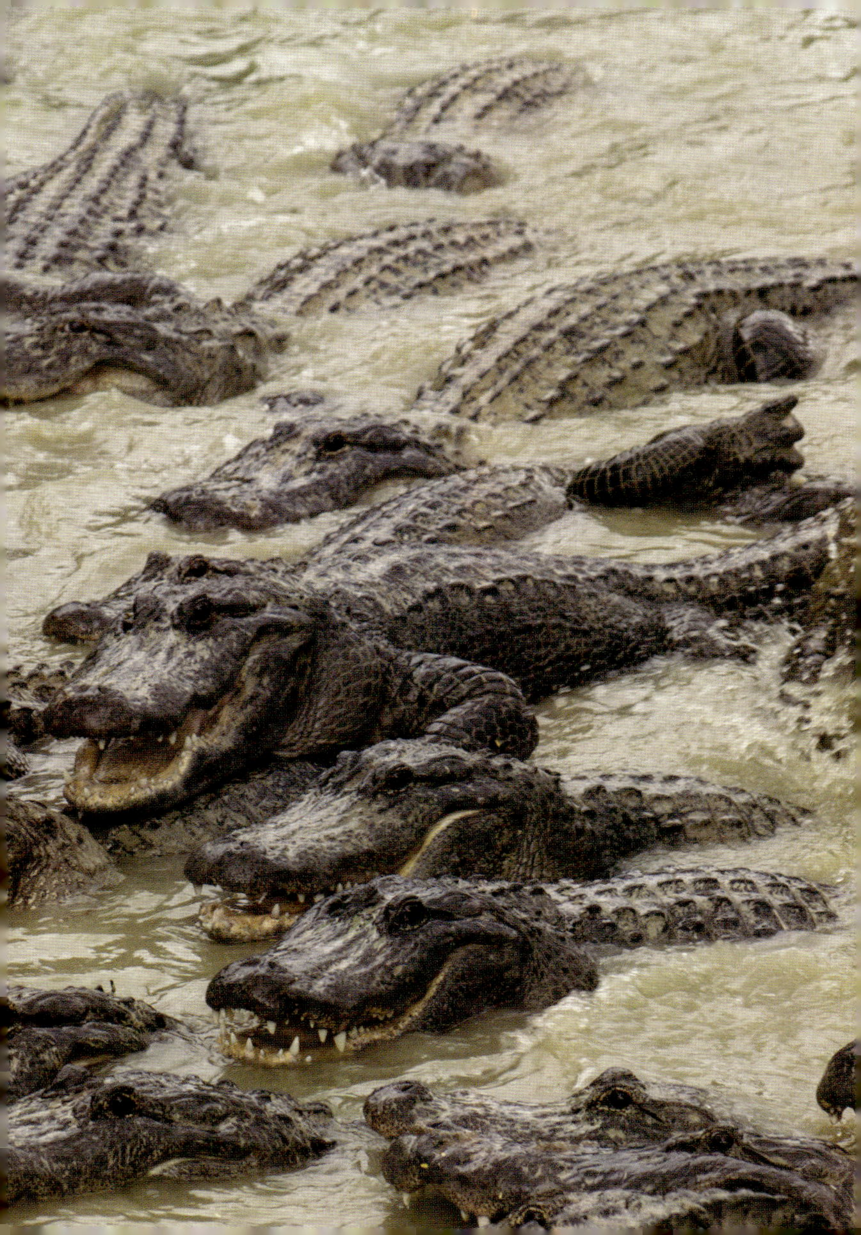

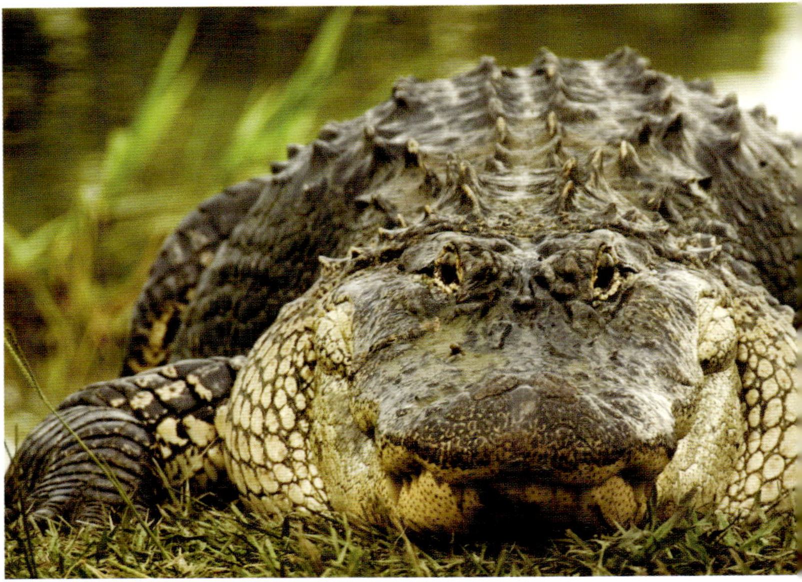

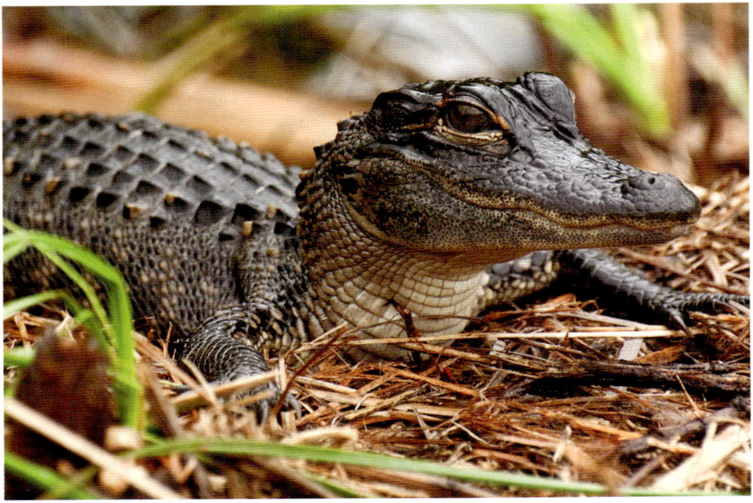

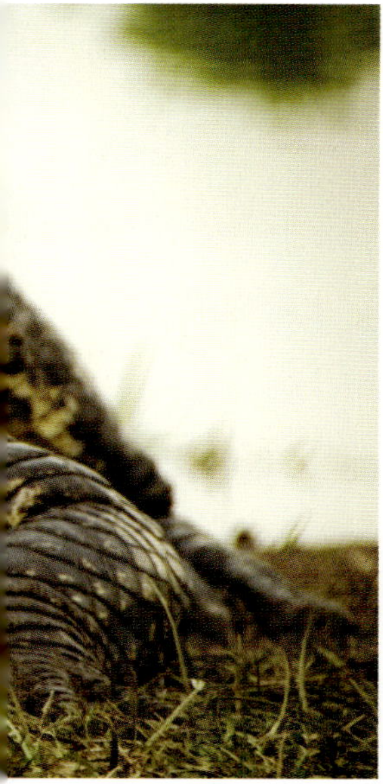

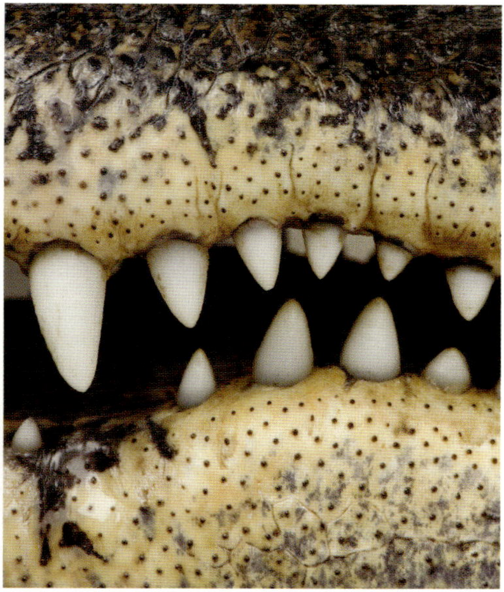

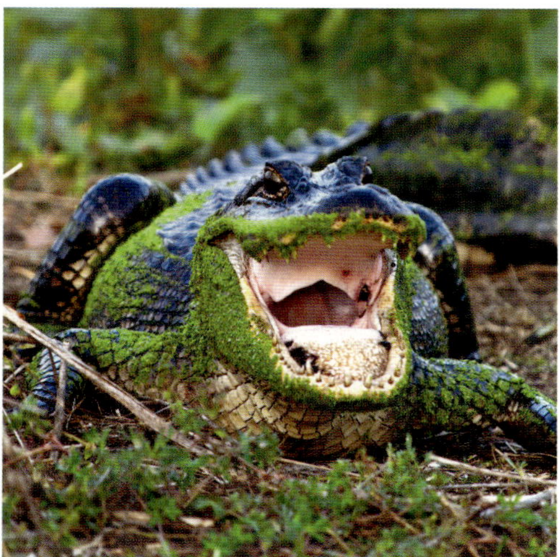

ALL PHOTOGRAPHS:

American alligator

Alligators can be separated from crocodiles by their broad snout and darker colouration. Their powerful jaws hold up to 80 sharp teeth, used for holding on to their prey, which comprises raccoons, small mammals, fish, crabs and birds. Adult males can reach over 4m (13ft) in length, and weigh over 350kg (770lb).

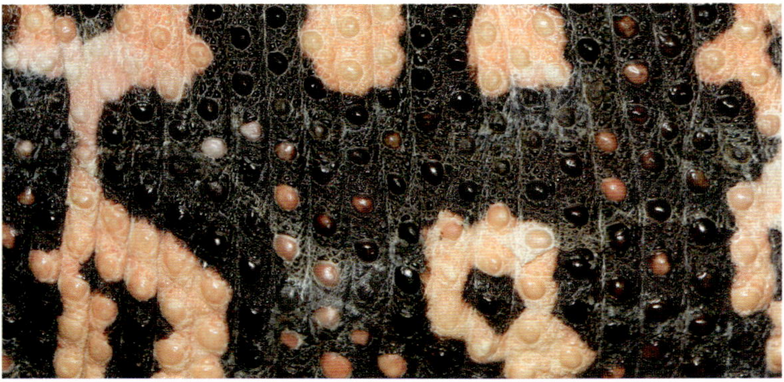

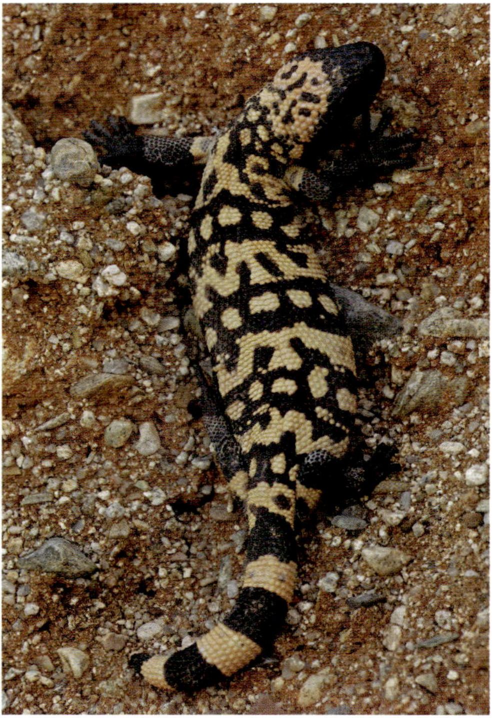

ALL PHOTOGRAPHS:

Gila monsters

The only venomous lizard native to the USA, the Gila monster lives in the deserts and scrublands of southwestern states. It has a studwork of small bones on its mottled pink and black hide, and is quite a squat, slow-moving lizard with a wide tail, which is used to store fat. Despite its fearsome reputation, it is quite harmless to humans.

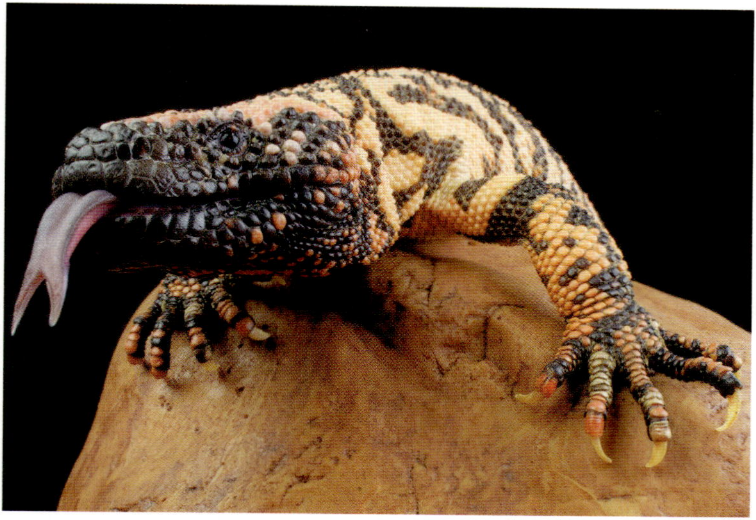

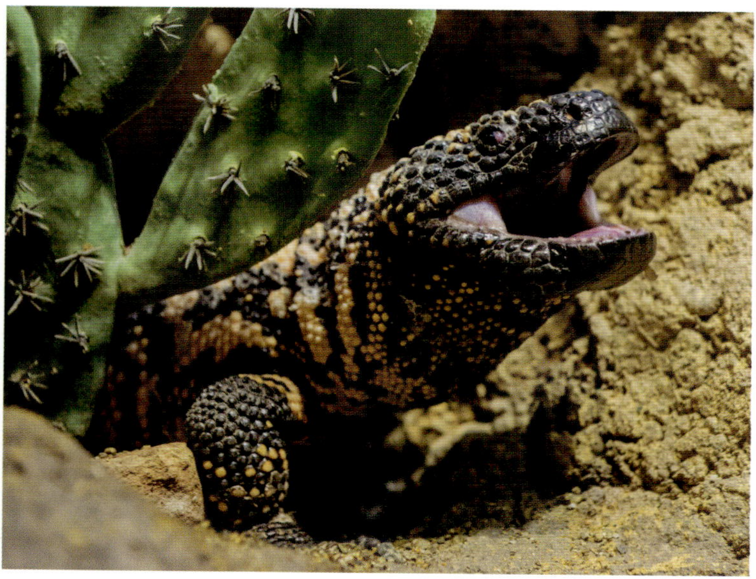

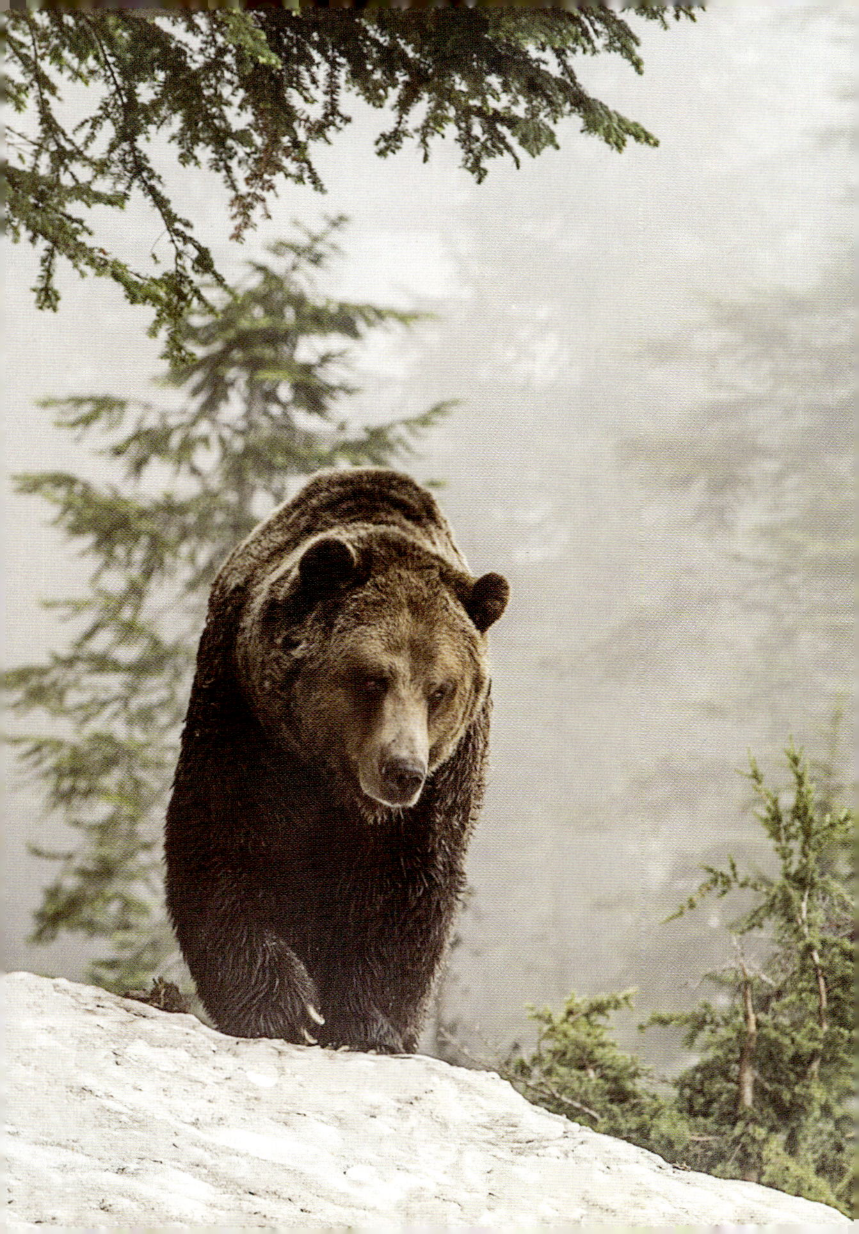

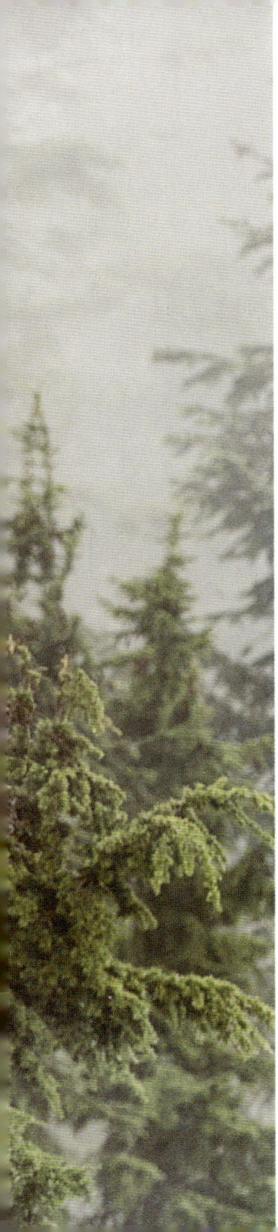

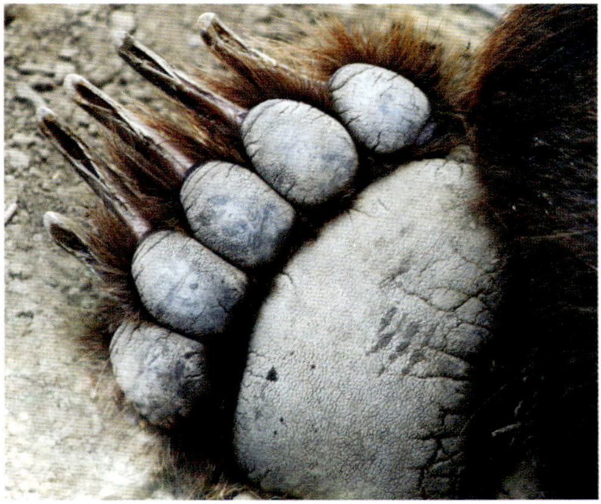

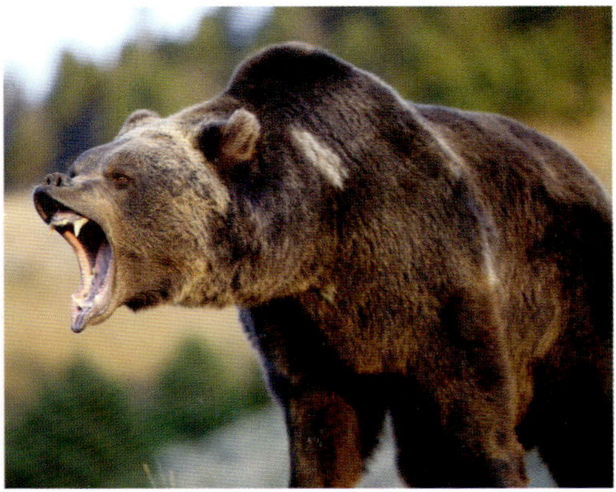

ALL PHOTOGRAPHS:

Grizzly bear

Also known as the North American brown bear, the term 'grizzly' relates to the blonde-tipped fur present on the backs of these impressive and extremely powerful animals. They inhabit Alaska, western Canada and some northwestern states of the United States. A grizzly bear's sharp claws, which it used to climb, dig and catch food, are 5–7cm (2–3in) long.

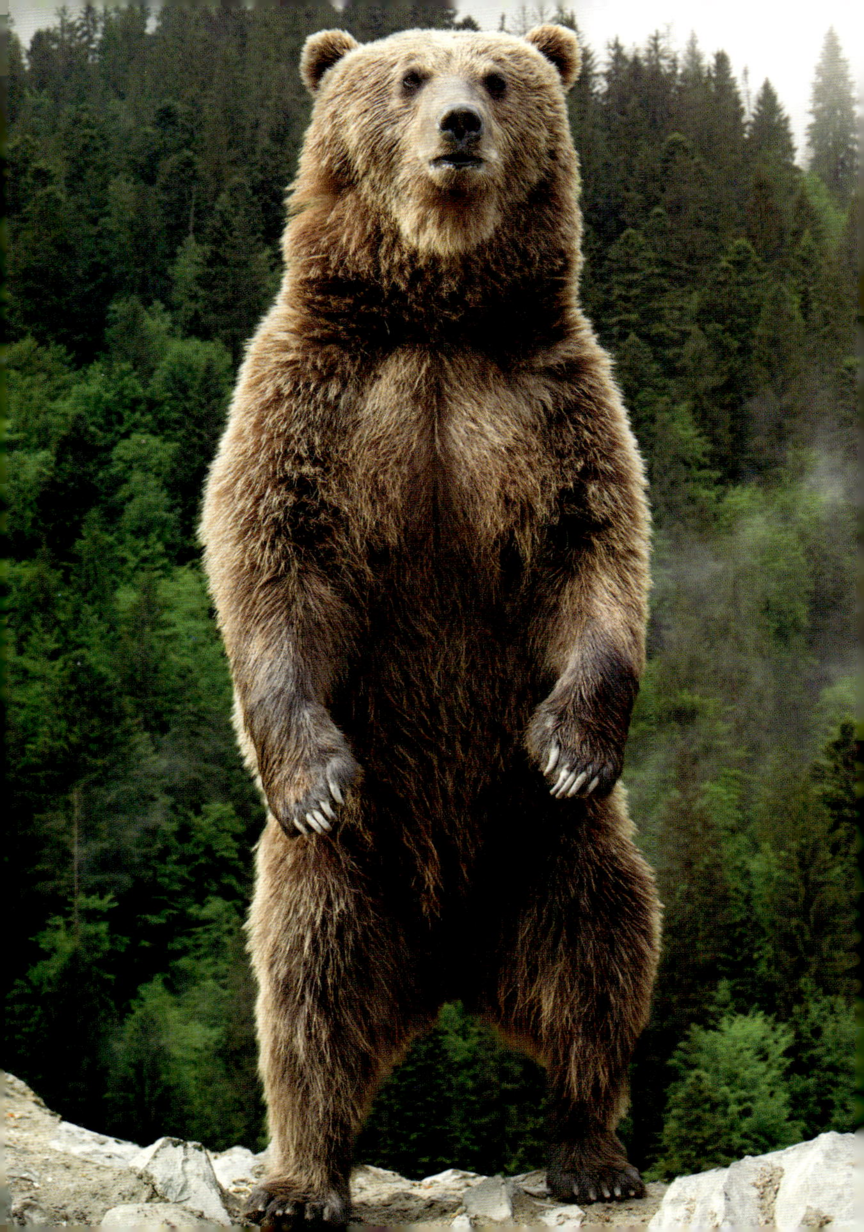

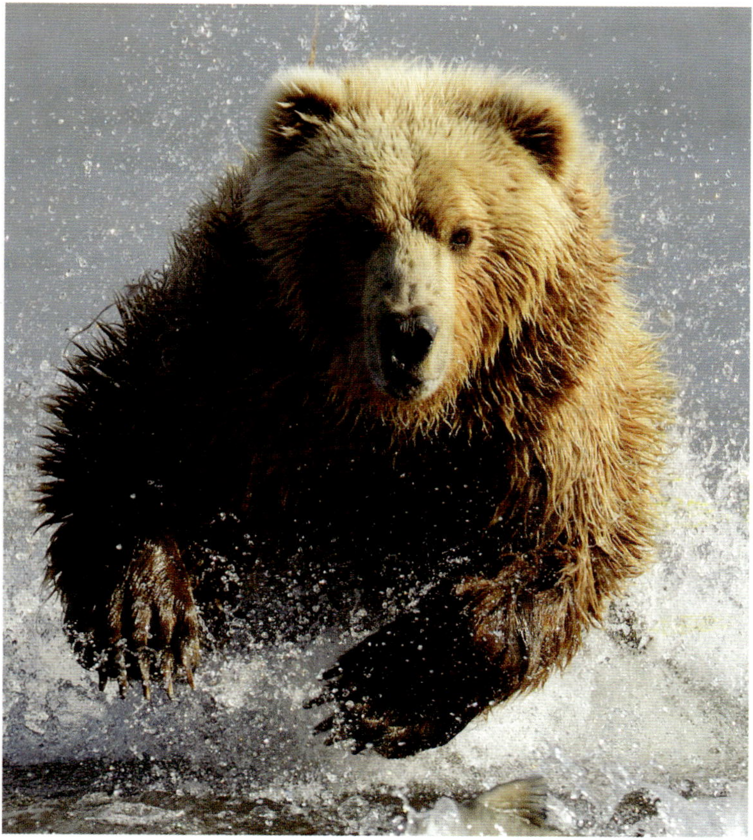

ALL PHOTOGRAPHS:

Grizzly bear

On average, grizzly bears reach a length of 2m (6.5ft), but they appear much larger when standing on their hind legs, something they do when they sense danger. They have acute senses of smell and hearing, but relatively poor sight. They avoid human contact, but if surprised, or protecting cubs, they can be very aggressive.

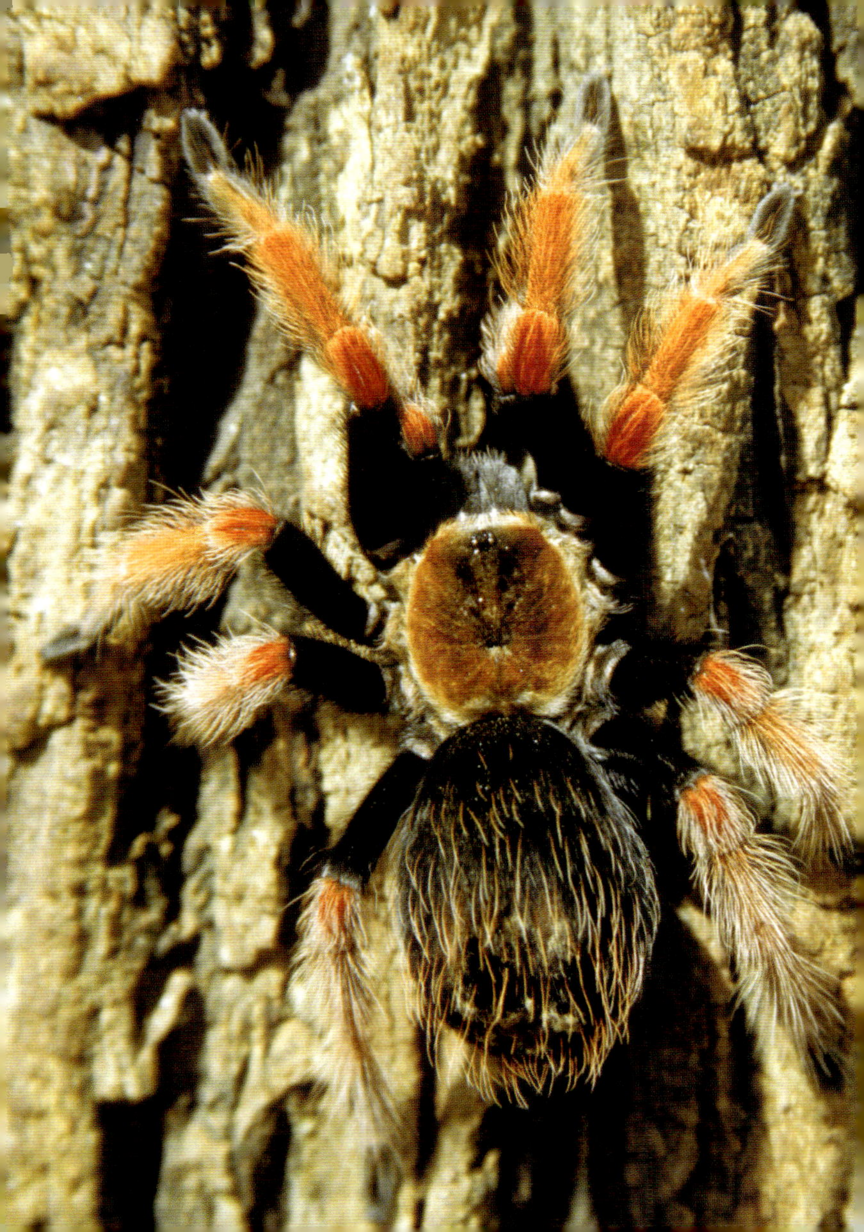

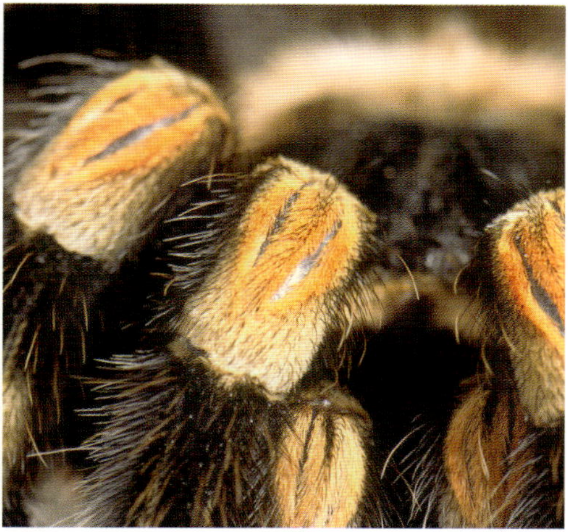

ALL PHOTOGRAPHS:

Mexican red-kneed spider

These tarantulas (there are two similar species), are so-called because of the orange-red striping on their legs. A native of Mexico, they have a body size of 10cm (4in) and a leg span of up to 15cm (6in). They feed on insects, crickets, small frogs and rodents, which they catch by active nocturnal hunting.

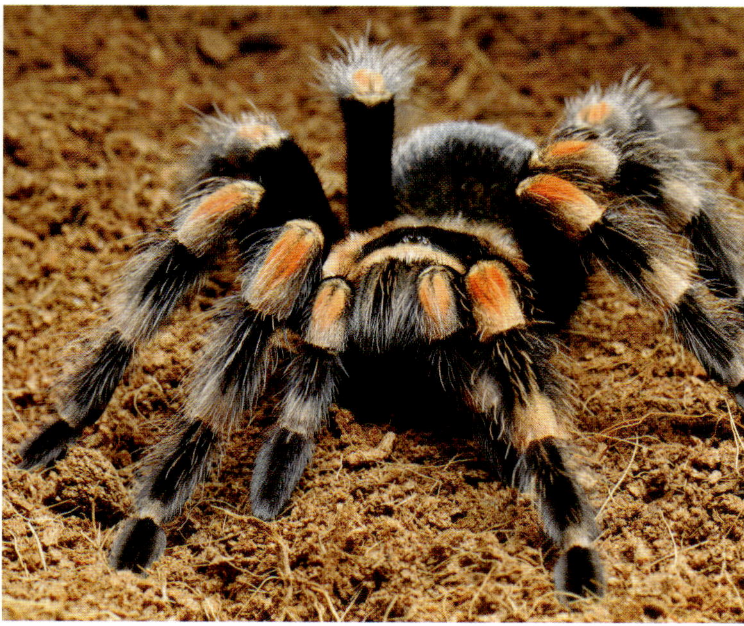

Female polar bear with cubs

Female polar bears give birth to an average litter of two cubs during the Arctic winter. The cubs are born blind and weaned in a den dug into a snow drift. In spring, the female will lead them across the pack ice in search of food.

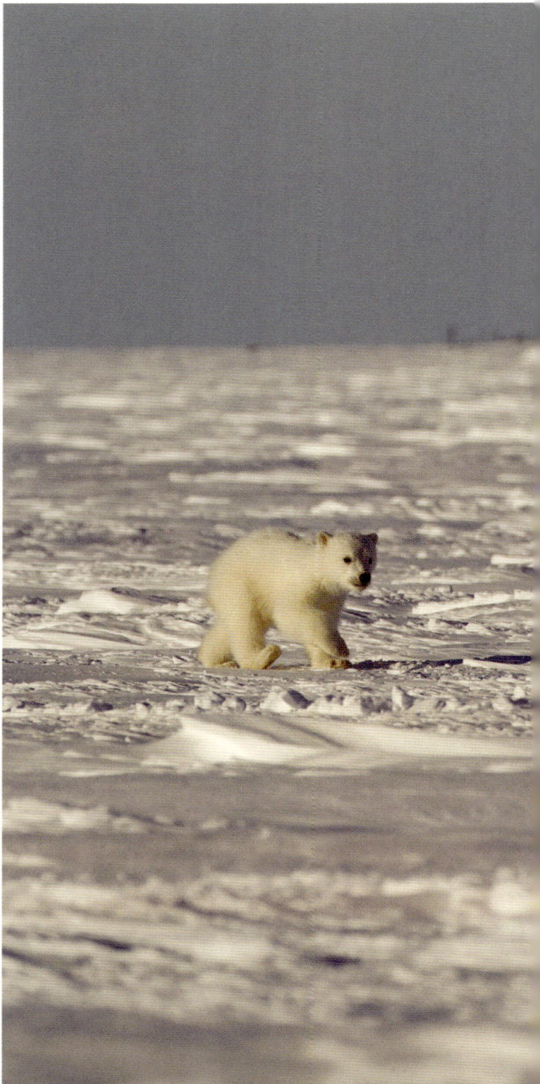

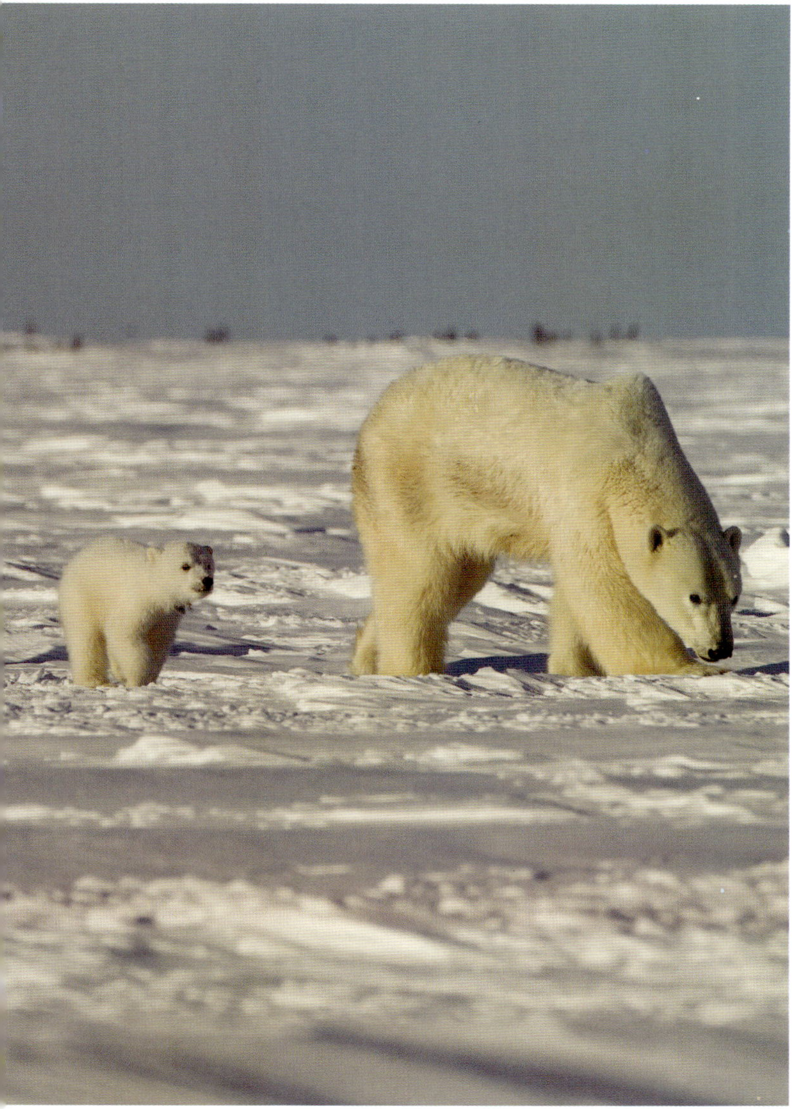

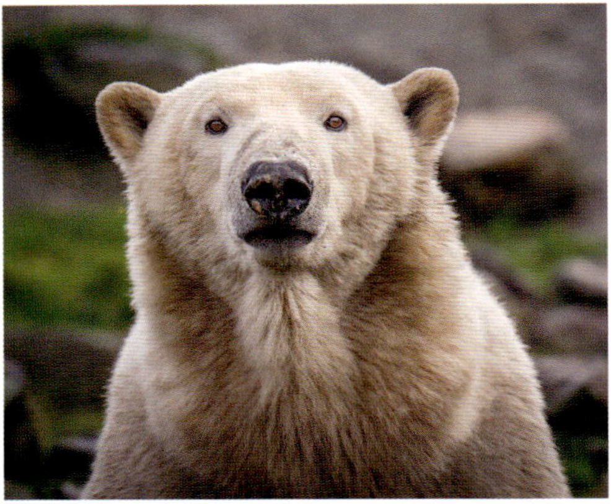

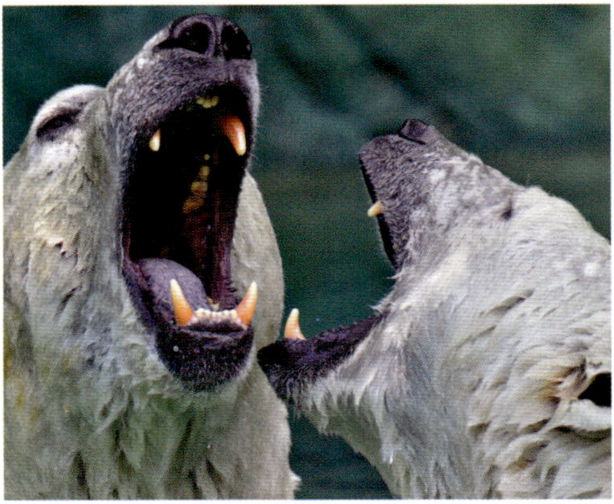

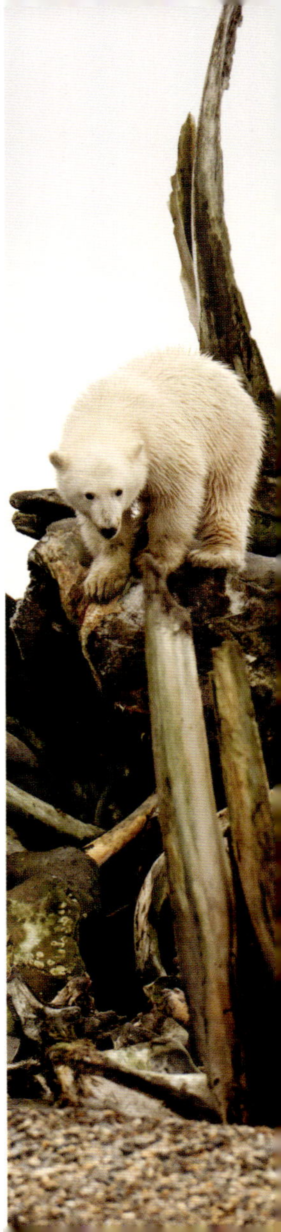

ALL PHOTOGRAPHS:

Polar bear

This is the largest species of bear, with adult males growing up to 3m (9.8ft). It is superbly adapted to a life foraging over the Arctic ice, having an insulating layer of thick white fur. It has an excellent sense of smell that allows it to detect seals, its primary prey, and carrion, such as this whale carcass, from a long way off.

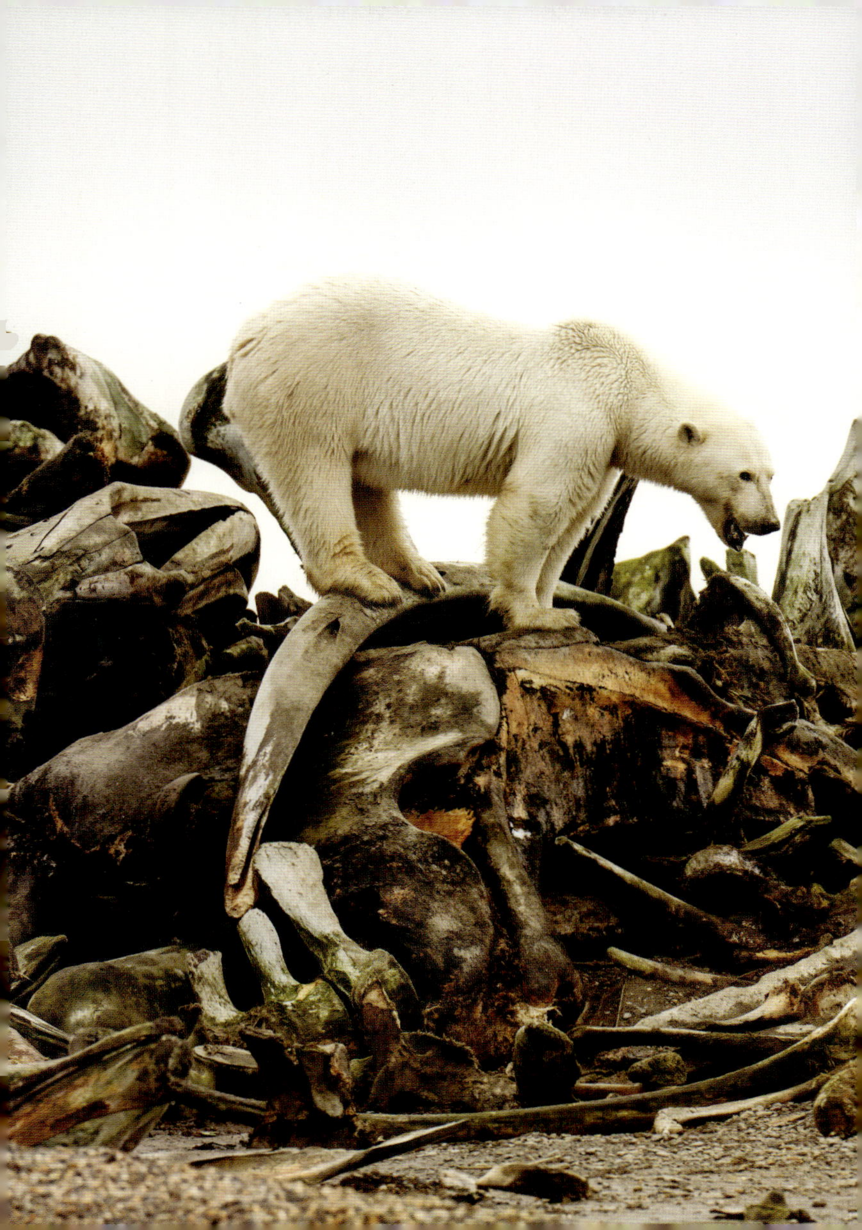

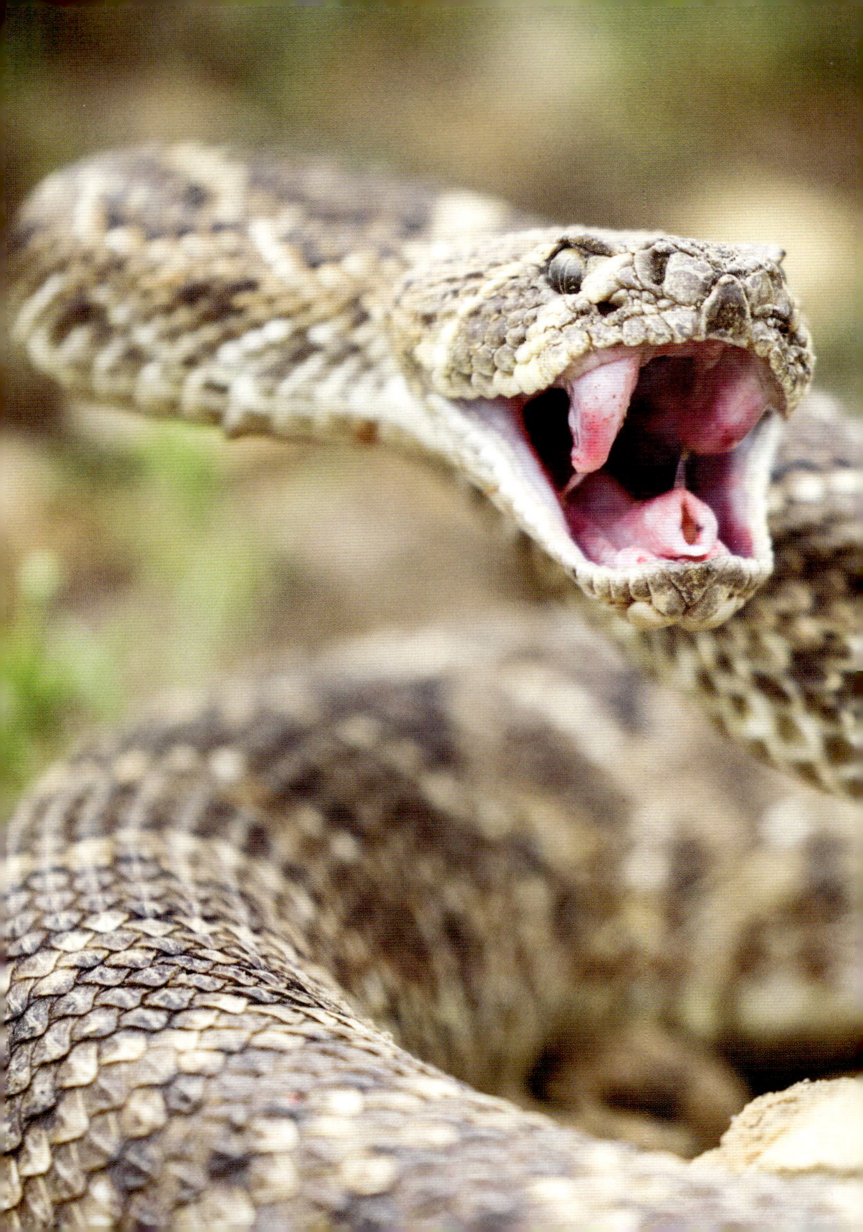

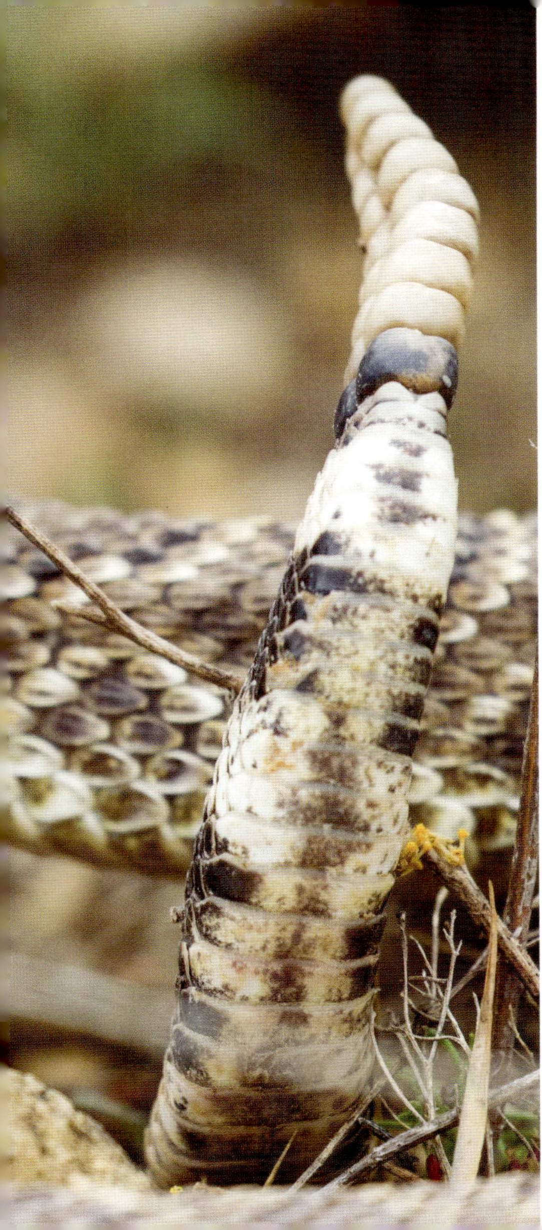

Western diamond rattlesnake

A species inhabiting the southwestern states of the United States, it can be identified by prominent black-and-white bands above the rattle at the end of its tail. It averages lengths of 1–1.5m (3.3–5ft), and uses heat-sensing receptors to track its prey of small mammals and birds. The retractable fangs measure 5cm (2in) and release a potent venom which attacks its victim's blood system.

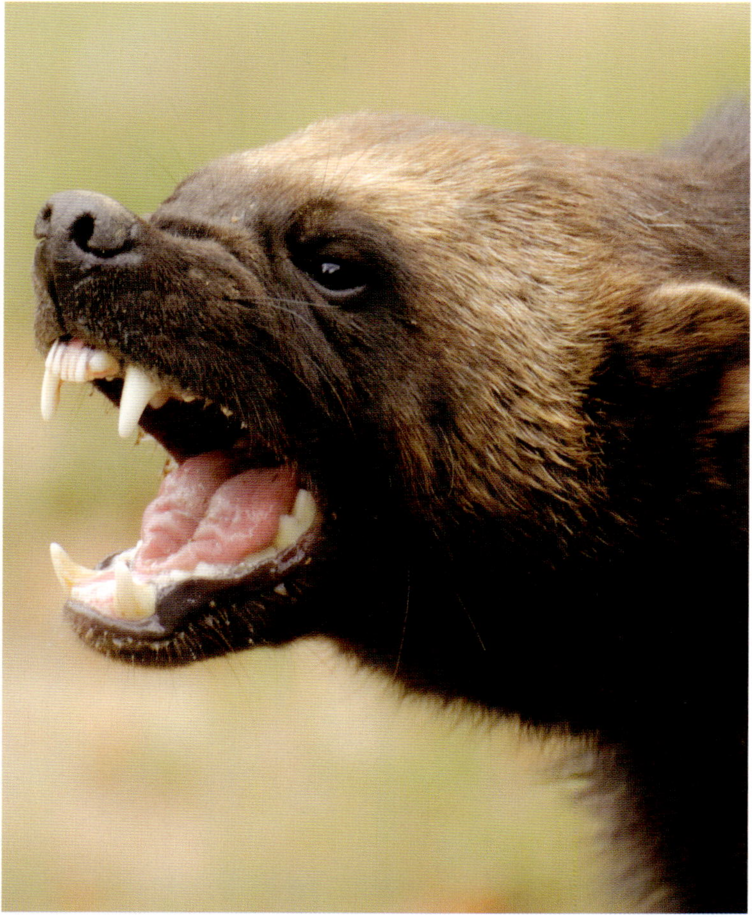

ALL PHOTOGRAPHS:
Wolverine
Wolverines are powerful animals with thick, muscley limbs and large paws that have strong claws. They have a coat of thick chocolate-brown fur and are as big as a medium-sized dog. Adept at climbing, and skilled hunters when they need to be, they obtain much of their food from scavenging carcases of animals killed by other predators.

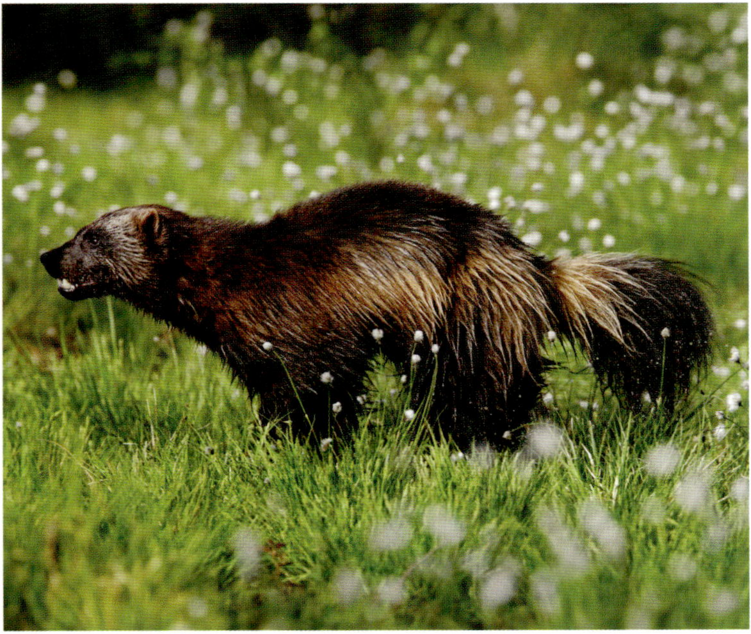

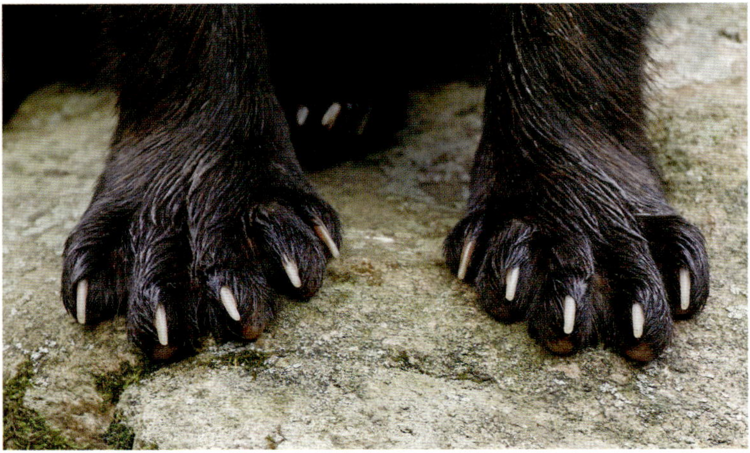

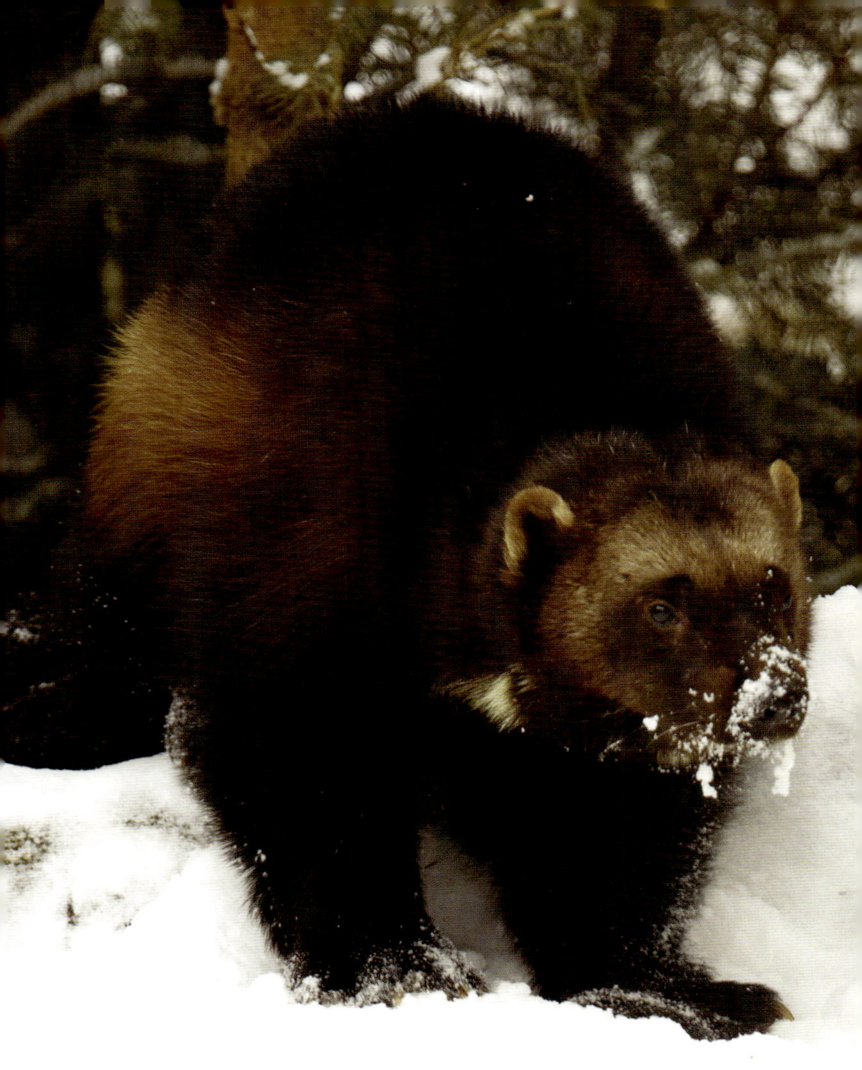

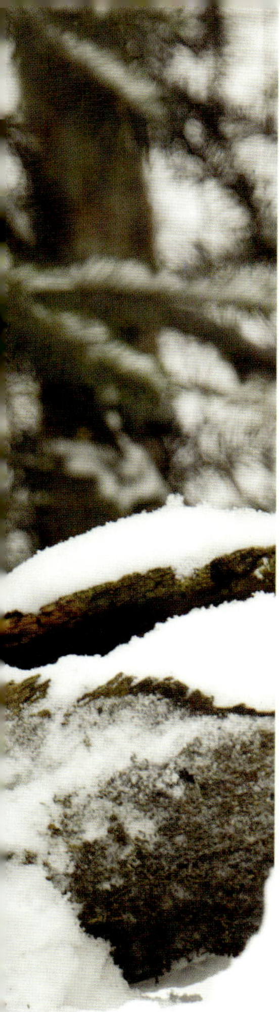

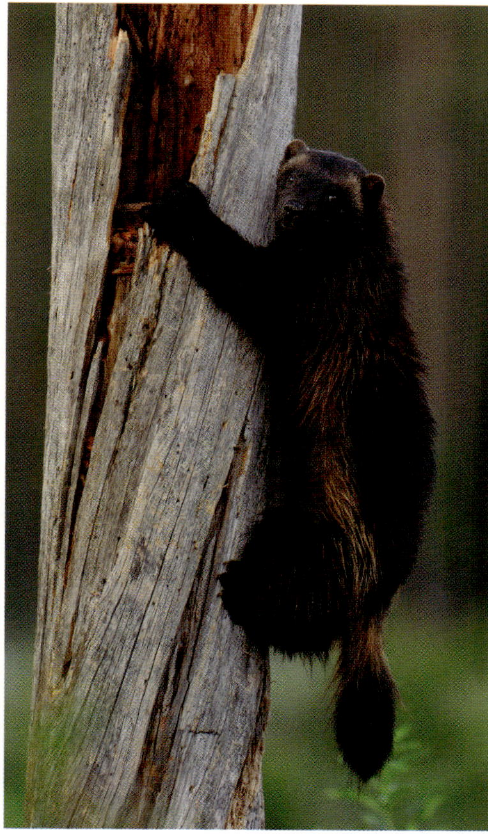

ALL PHOTOGRAPHS:

Wolverine

Wolverines have a widespread distribution around the Arctic regions of North America, being well adapted to survive in this inhospitable climate. They have earned a fearsome reputation, but rarely come into contact with humans. In fact, their fur, which is resistant to frost, is prized amongst hunters, who take a toll on the population.

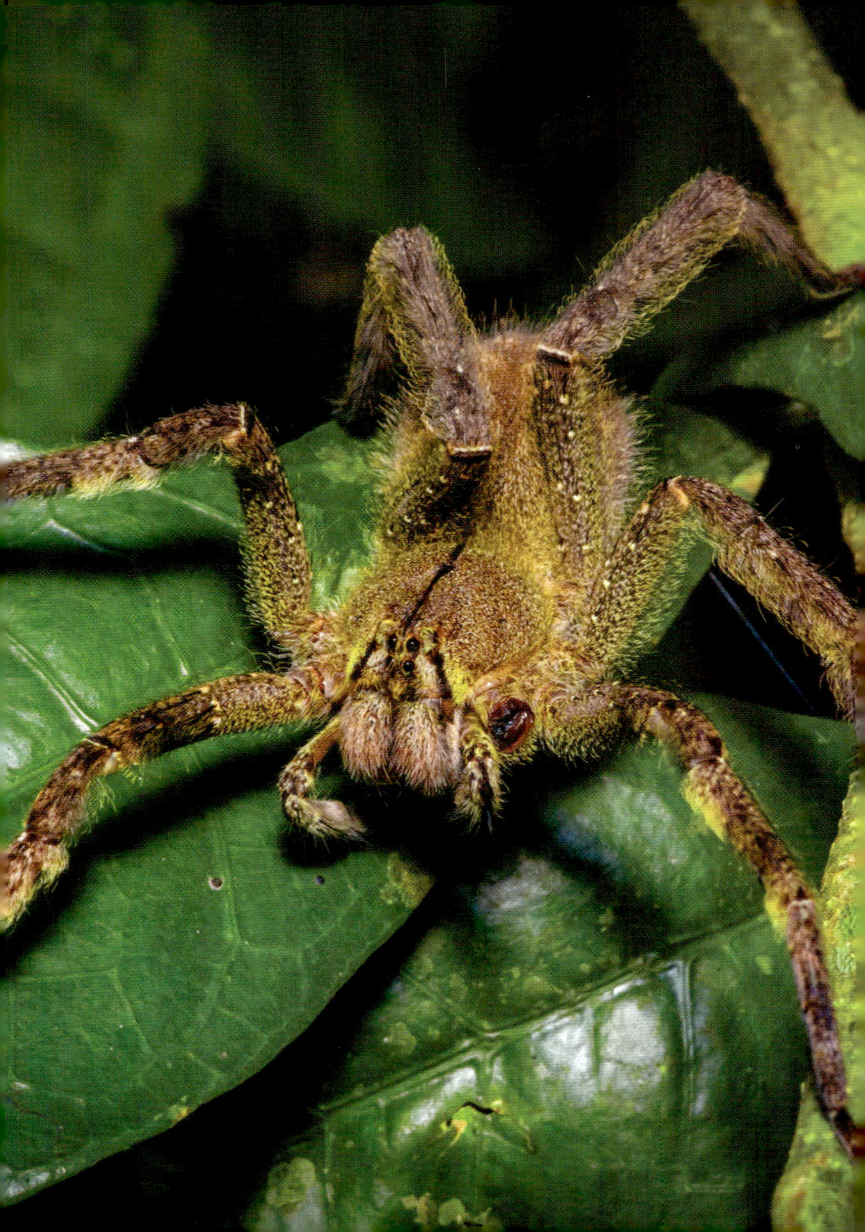

South and Central America

From high, snowcapped Andean mountains, to wind-blasted, hostile deserts; and from lush, vibrant rainforest cut by the mighty Amazon to glaciers emptying the snowfall of centuries past into crystal clear lakes the colour of opals, the South and Central America explodes with dramatic scenery and abundant wildlife. Rainbow-hued parrots and macaws call it home, condors and vultures patrol the skies, while its forests hold secrets still awaiting discovery. Pumas hunt the high, windswept Patagonian steppe, and jaguars stalk unchallenged in thick Amazonian jungle. Colourful butterflies dance in the breeze, hummingbirds dart and squabble like so many outsize bees, agile monkeys climb through the tree canopy; sloths crawl, harpy eagles strike. It's such a vibrant, captivating and richly diverse land. The human element should not be ignored, providing a rich cultural heritage with breathtaking legacy of architecture and monument. The legacy of ancient civilisations have left their obvious mark on the landscape, but even now some indigenous tribes live harmoniously with nature in remote isolation.

This chapter will reveal some of South and Central America's most amazing predatory animals, showcasing their special features and highlighting their particular skills. A selection of the continent's most impressive reptiles, arachnids, insects, birds, fish and mammals are featured. Some large, others small; some stealthy hunters, others opportunistic feeders.

All deadly.

OPPOSITE:

Brazilian wandering spider

A species commonly found in densely populated areas of southeastern Brazil, this spider has a body length of up to 5cm (2in) and a leg span of 15cm (6in).

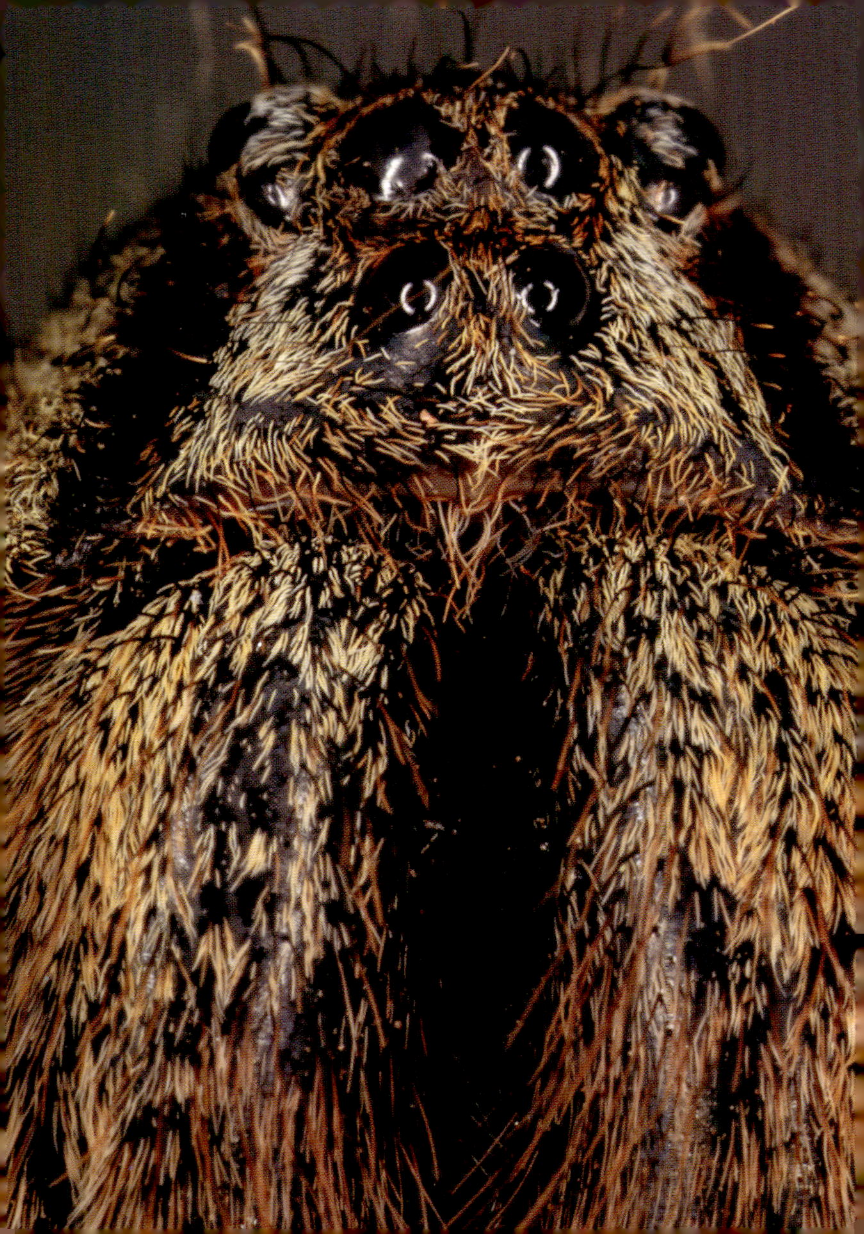

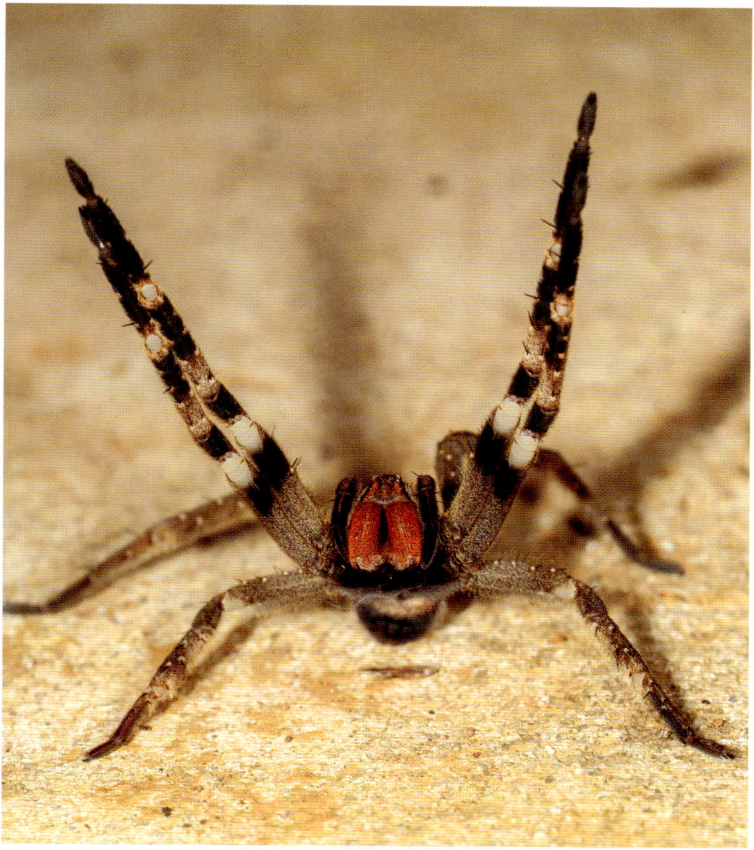

ALL PHOTOGRAPHS:
Brazilian wandering spider
These eight-eyed spiders have a body covered with thick brown hair. When threatened, they hold their front legs high, revealing an identifying black-and-white underside pattern, together with red jaws. Their bite can be painful, and on rare occasions fatal, although they often administer a 'dry bite' without releasing venom.

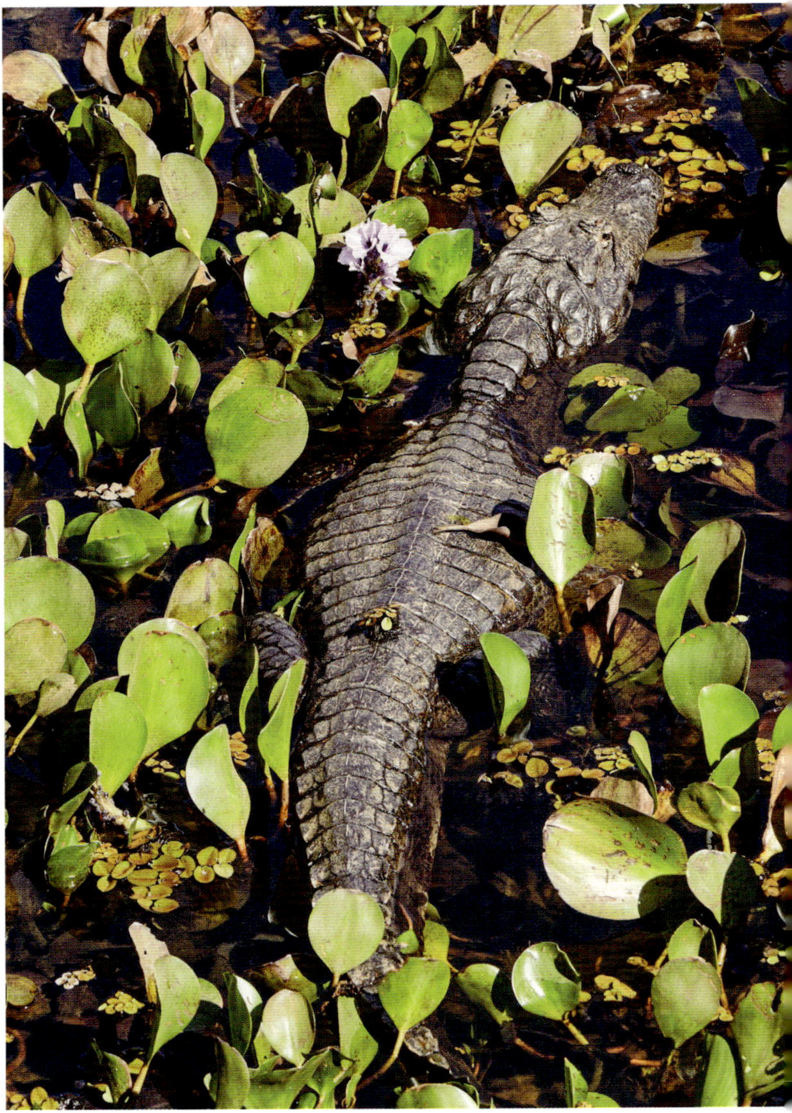

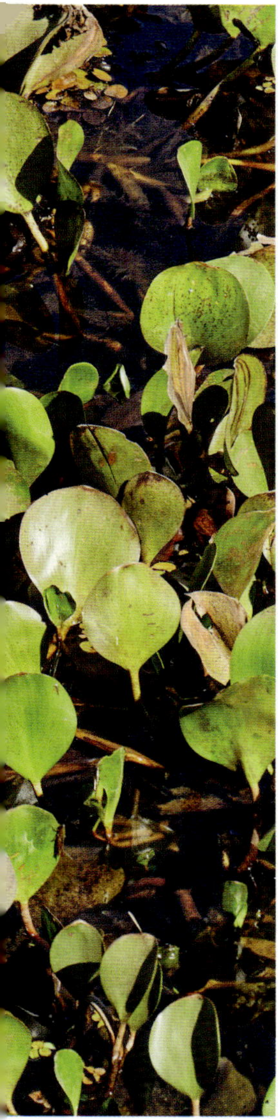

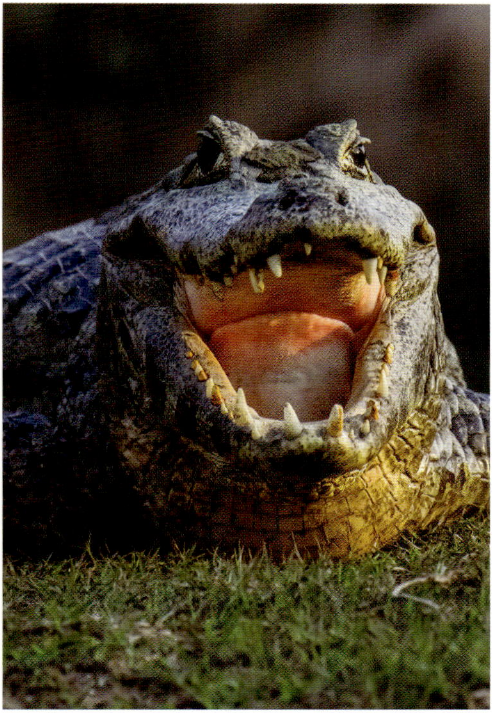

ALL PHOTOGRAPHS:

Black caiman

An almost-black crocodilian growing to 5m (16.5ft), this caiman inhabits wetland areas across the continent. It feeds on mammals, other reptiles, birds and fish, which are ambushed under the cover of darkness, when its stealthy movements and dark hide makes it almost invisible to its intended victim.

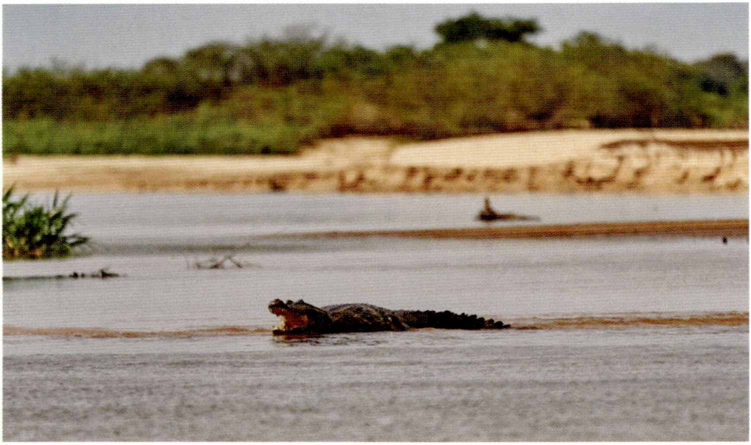

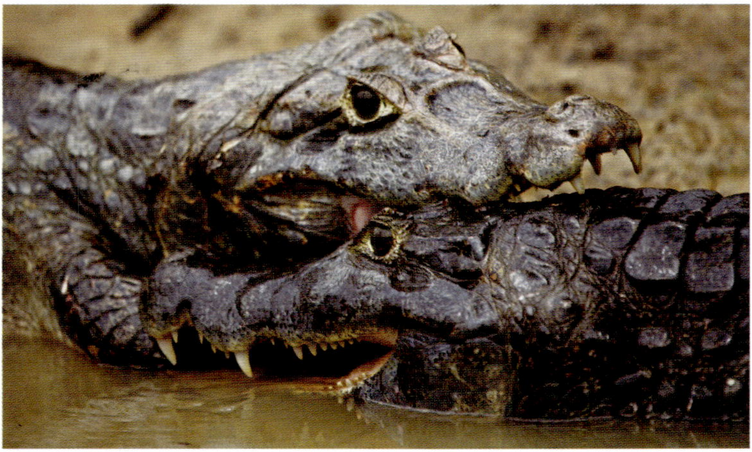

TOP AND BOTTOM:

Black caiman

These cold-blooded animals spend a lot of time basking in the sun with their mouths wide open. They have been mercilessly hunted in the past for their skins, suffering a serious depletion in numbers during the 20th century. Happily, numbers are recovering, although continued habitat destruction poses further long-term threats to the species.

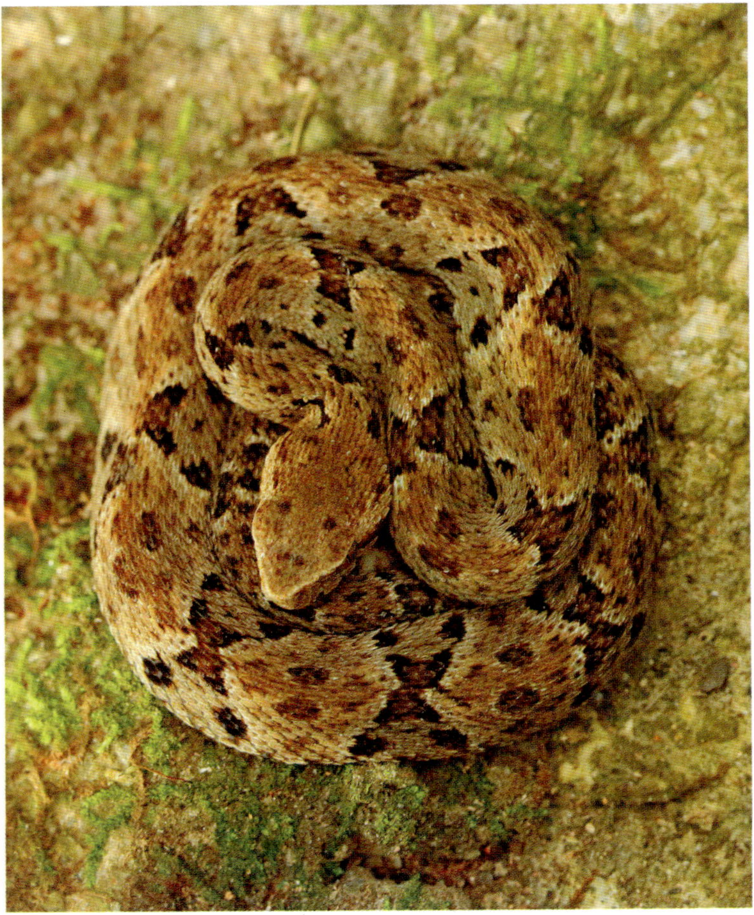

ABOVE:

Common lancehead snake

'Lancehead', or commonly the French 'fer-de-lance', is a reference to the snake's pointed head shape. Found in tropical lowlands east of the Andes, it is coloured in various shades of brown, green and grey, which provides excellent camouflage for this 120–150cm (4–5ft) predator of small rodents, frogs and lizards.

Common lancehead snake

The species is quite common, hunts at night and has golden-yellow eyes. It will wait motionless for prey in areas frequented by people, such as plantations, where unwary workers can get bitten. The venom impairs coagulation of the blood, and although painful and inhibiting, is rarely fatal to humans.

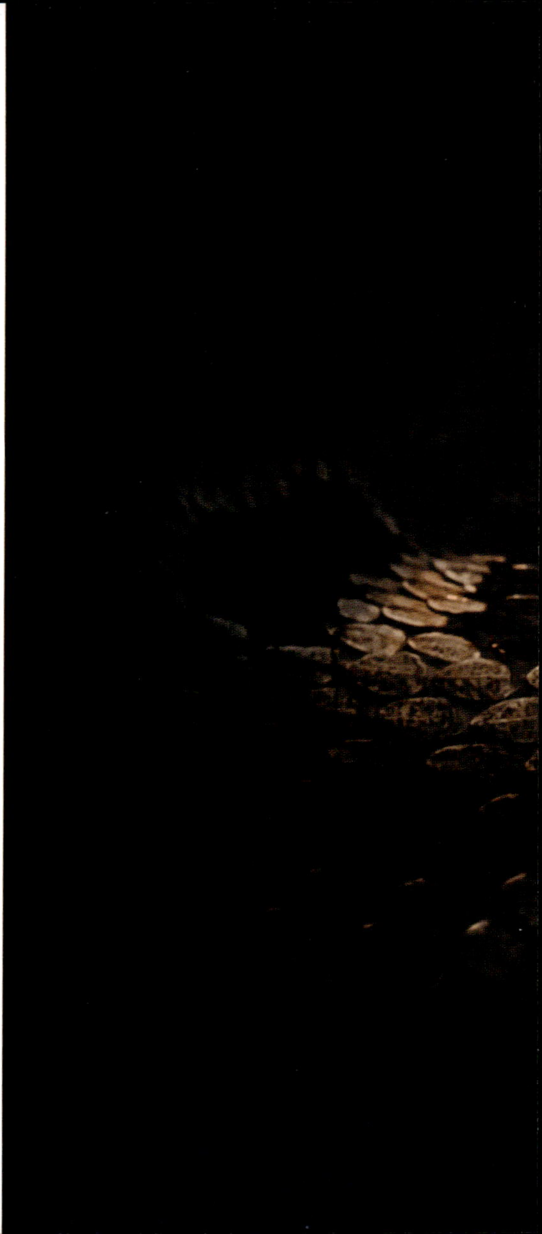

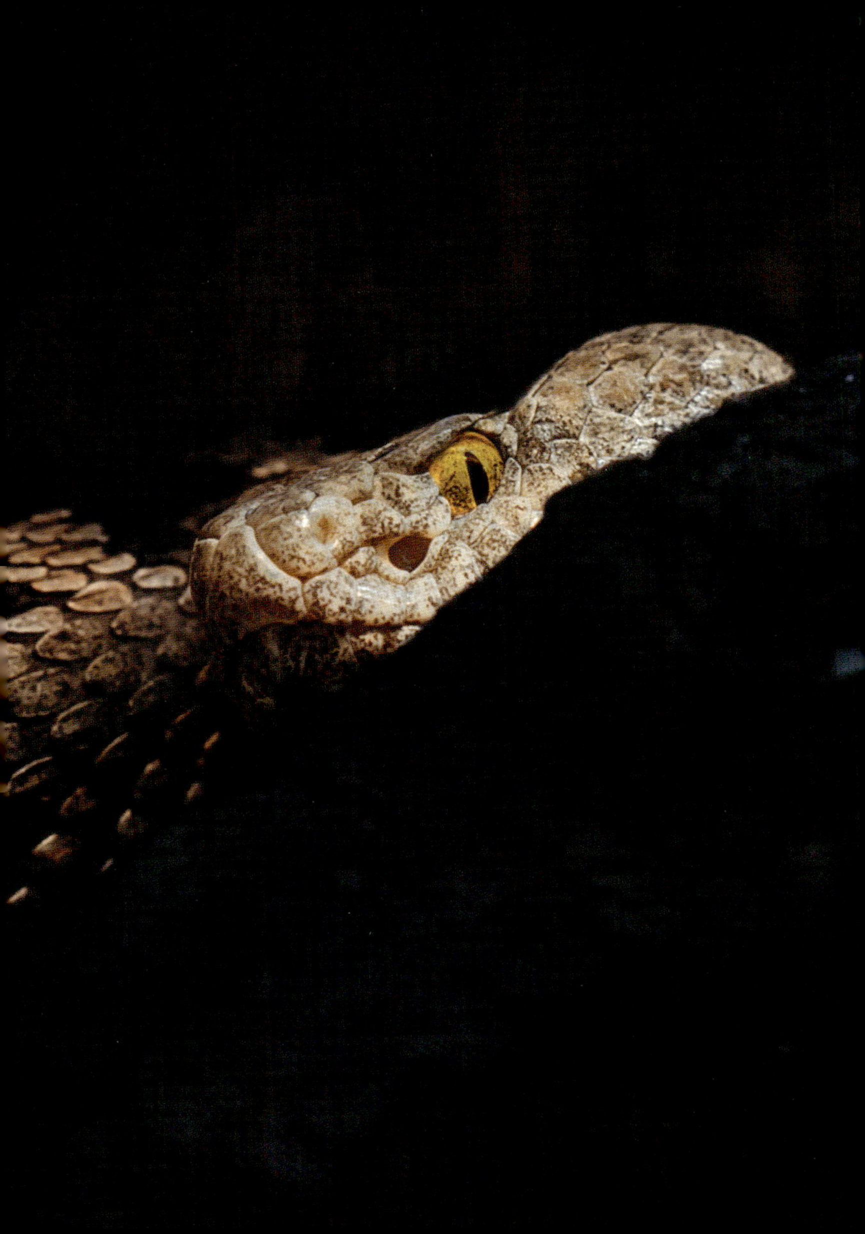

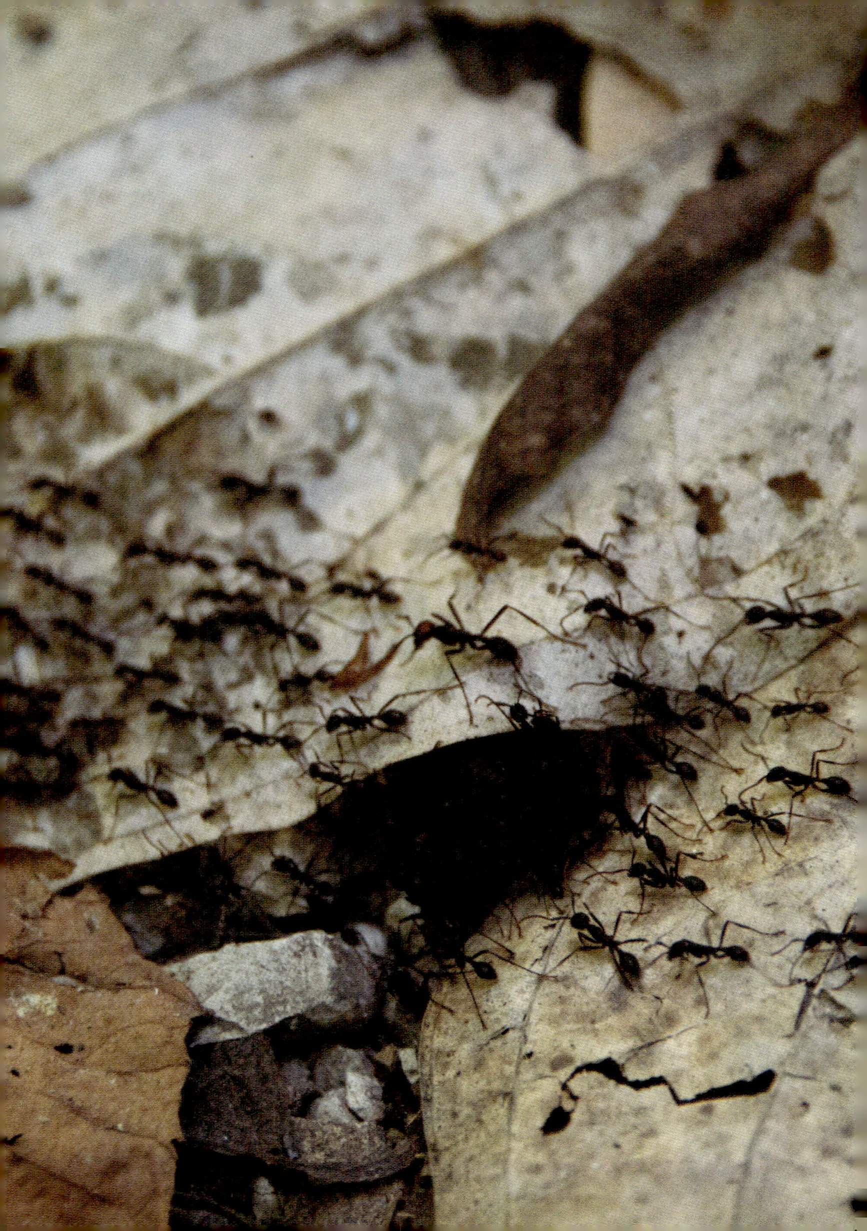

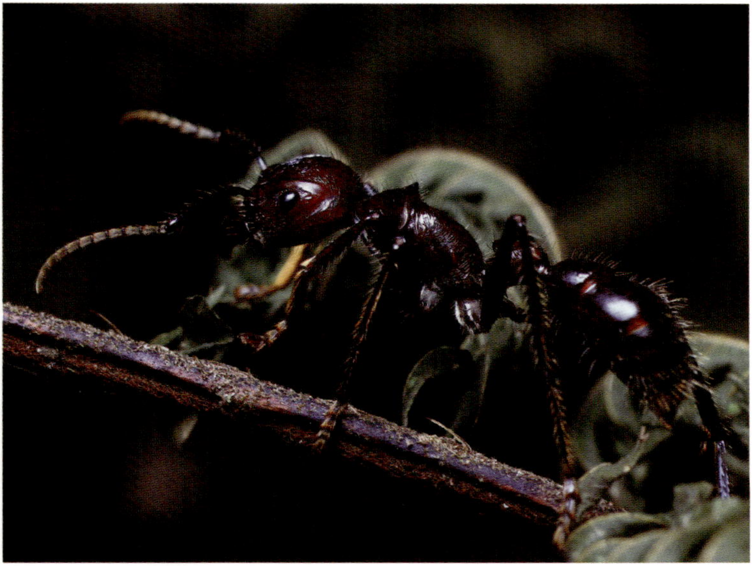

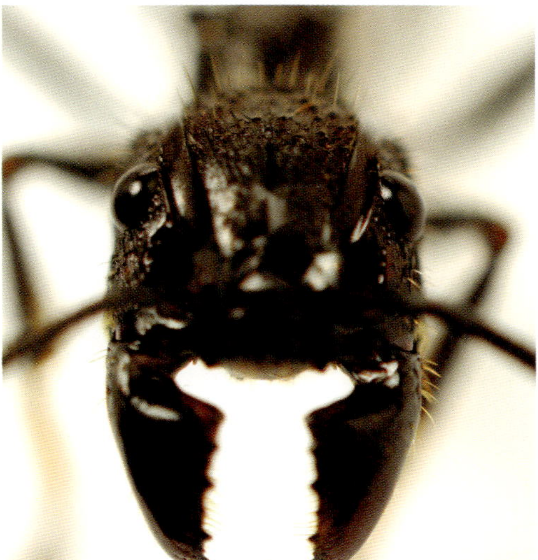

ALL PHOTOGRAPHS:

Bullet ant

A large, reddish-brown ant reaching 3cm (1.2in) in length, it is named after its sting, which is extremely painful and likened to being shot. The effect of the sting can last for 24 hours. These ants live in colonies sited at the base of trees, and forage amongst foliage for nectar and small invertebrates.

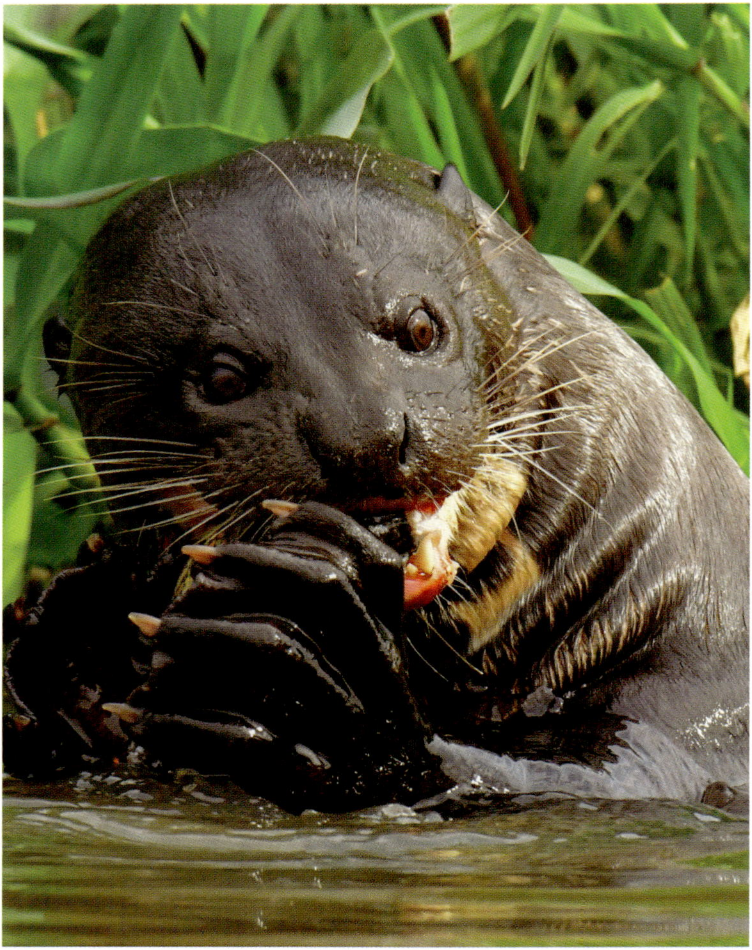

ALL PHOTOGRAPHS:
Giant otter
Up to 1.8m (6ft) long, this is the largest species of the mustelid family. It occupies freshwater areas of the Amazon basin and Pantanal where it lives in small family groups. Its strong jaws and sharp teeth allows it to feed on fish, crabs, turtles and small reptiles.

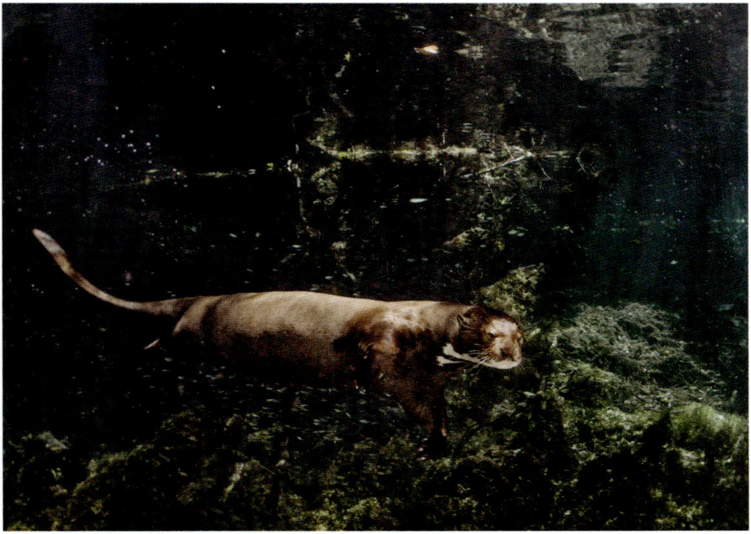

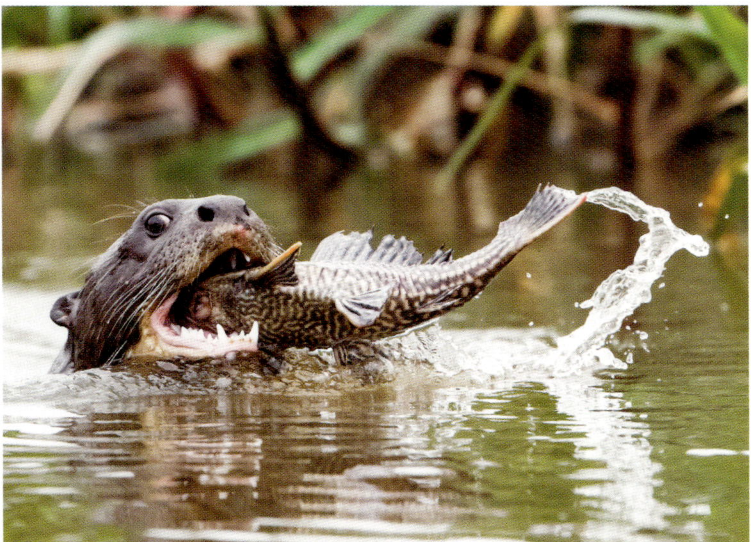

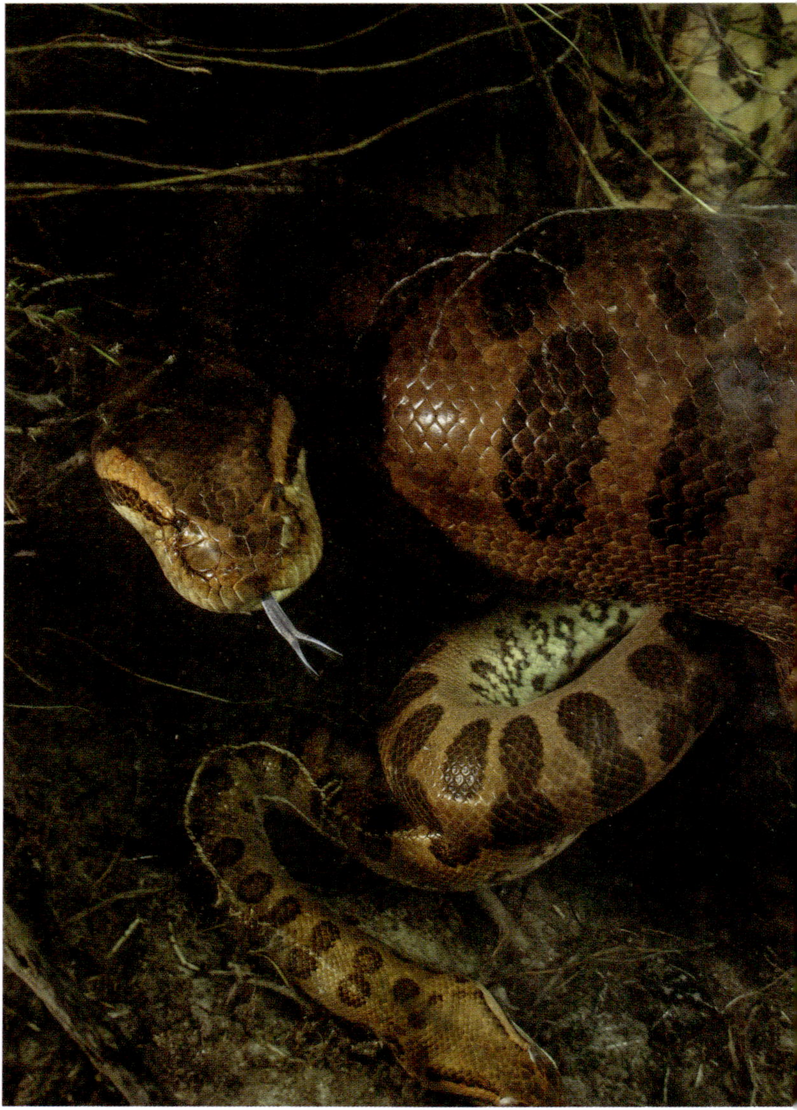

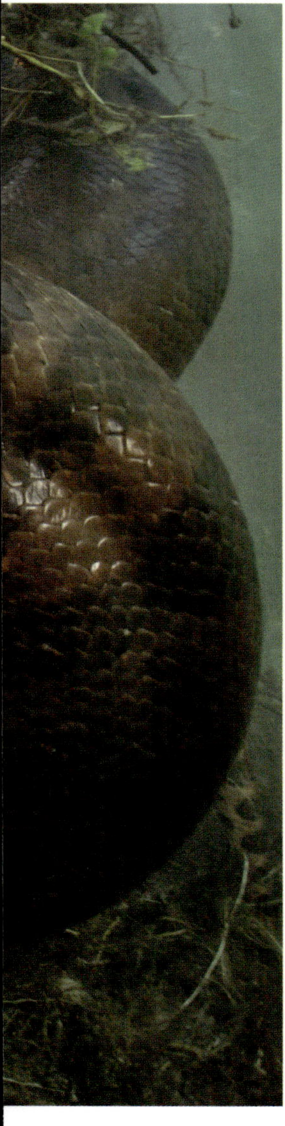

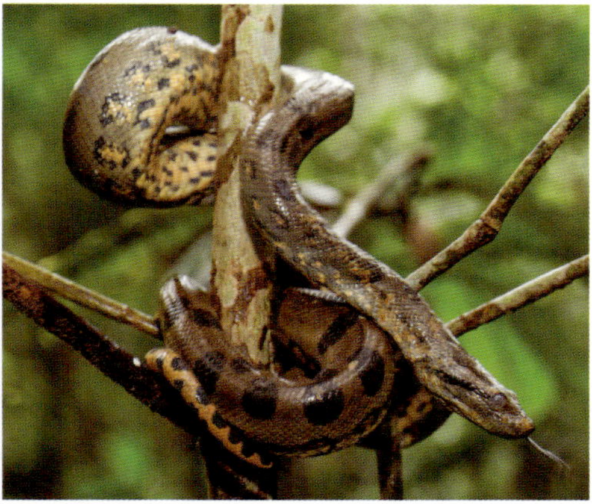

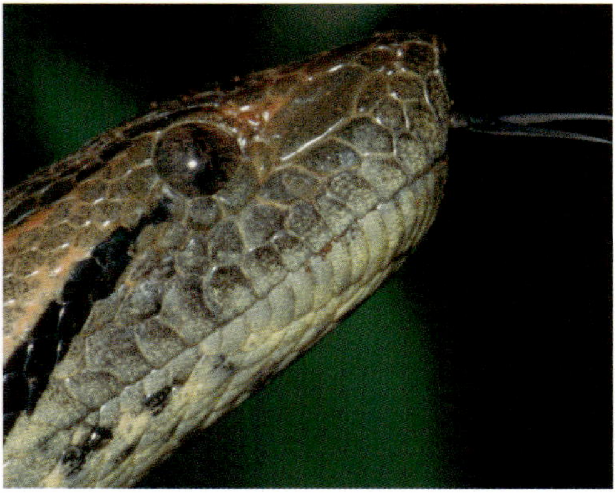

ALL PHOTOGRAPHS:

Green anaconda

So-called because of its green-coloured skin, this is one of the world's longest snakes, with males averaging 5m (16.4ft). It is mainly an aquatic hunter, preying on any suitably sized creature it can ambush as it stops to drink or feed. Anacondas are constrictors, grabbing prey in their jaws before suffocating the victim in its powerful coils.

Golden dart frog

Although only 5cm (2in) in size, this bright yellow, pale green or orange amphibian is the largest of the poison-dart frogs, and is considered to be the most poisonous animal in the world. It is endemic to the humid rainforests of Columbia where it hunts the forest floor for beetles, ants and other invertebrates.

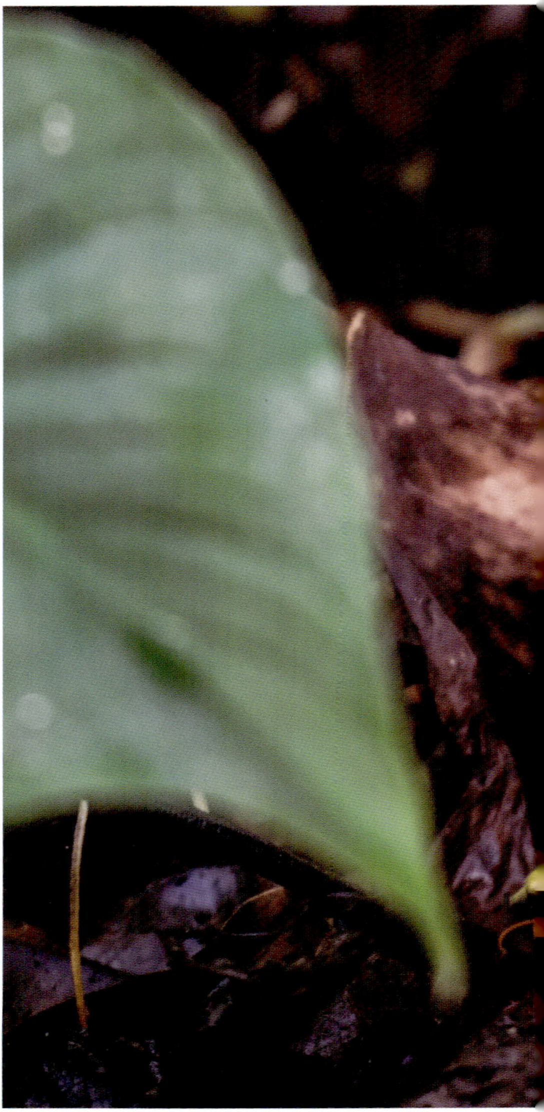

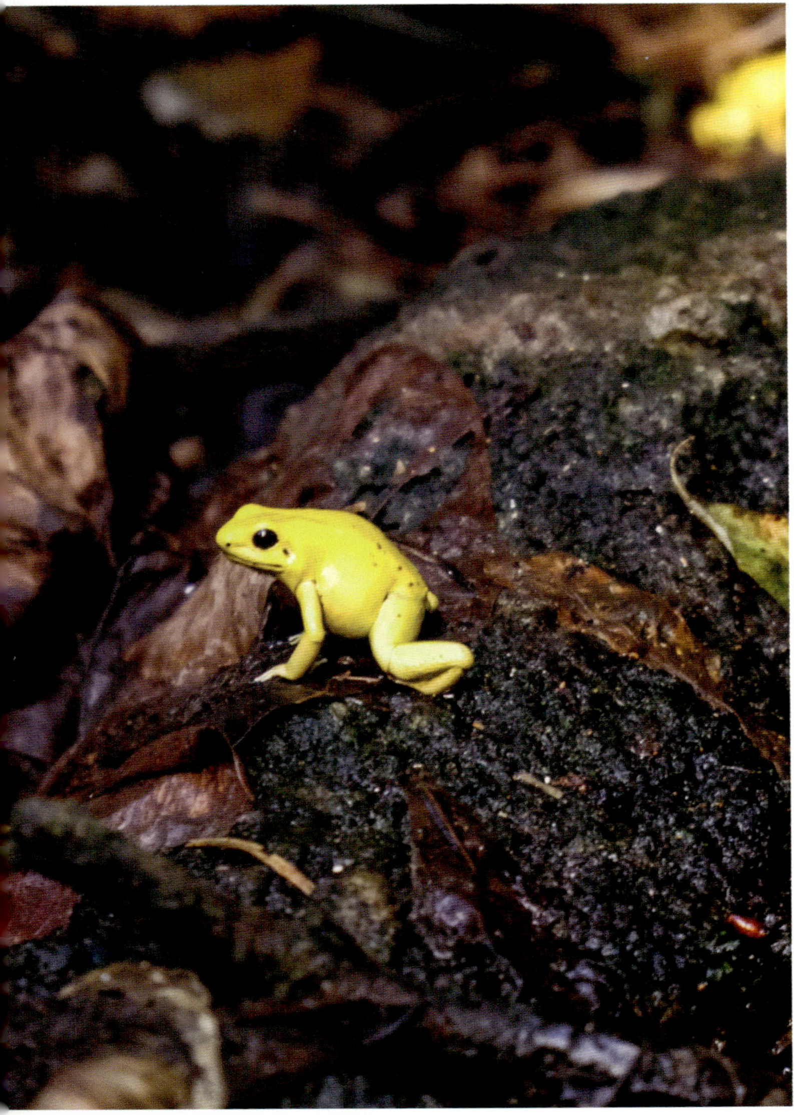

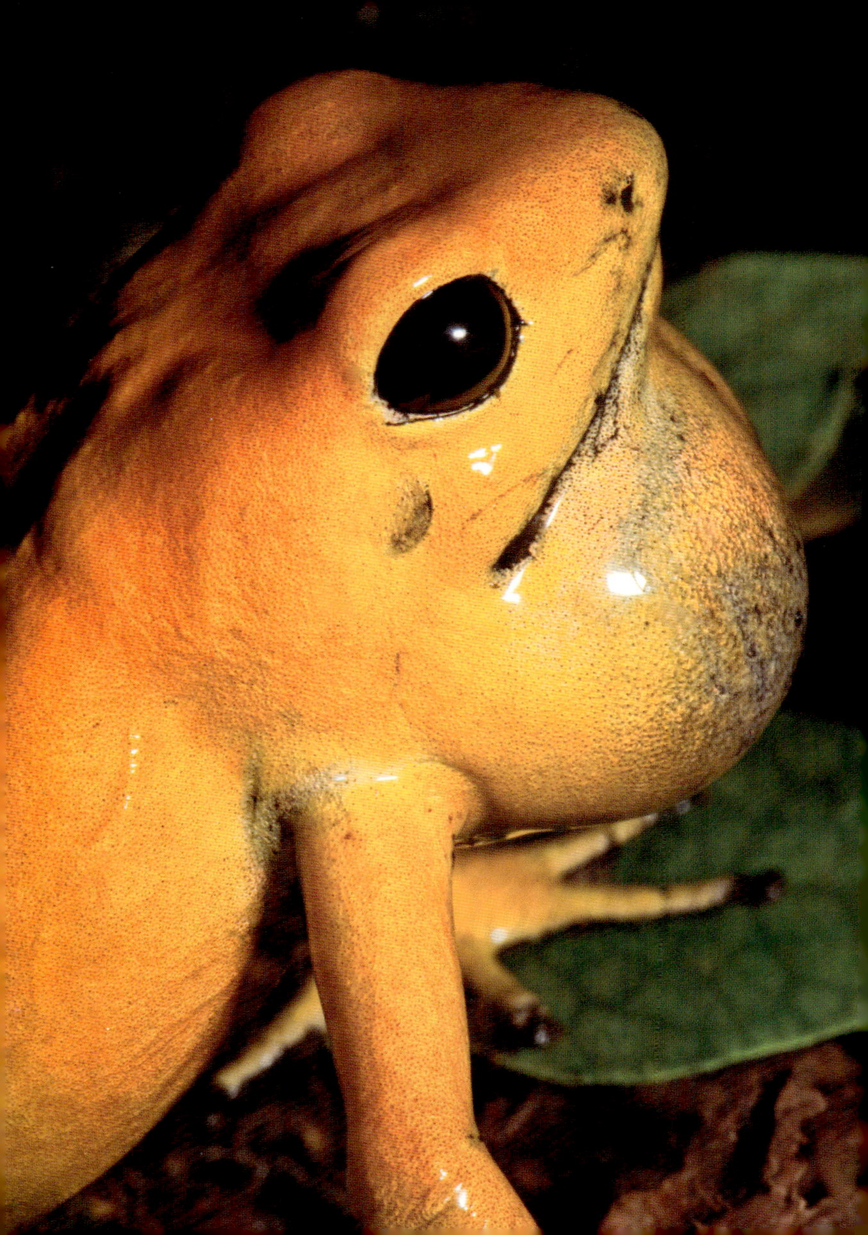

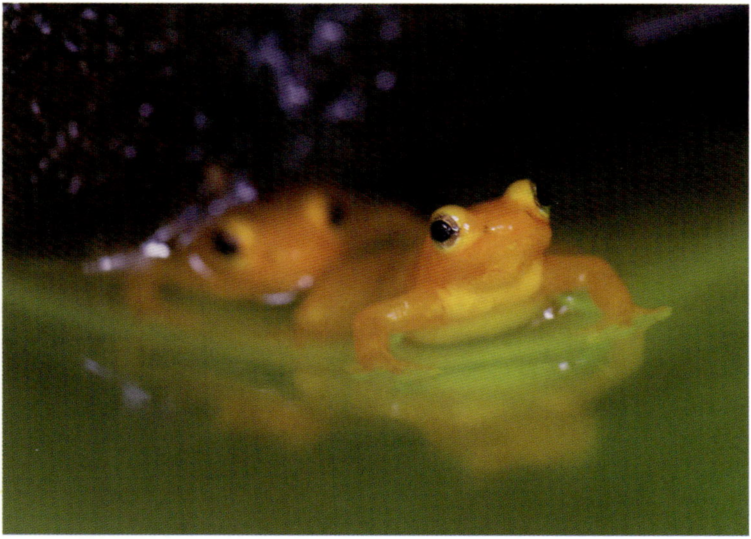

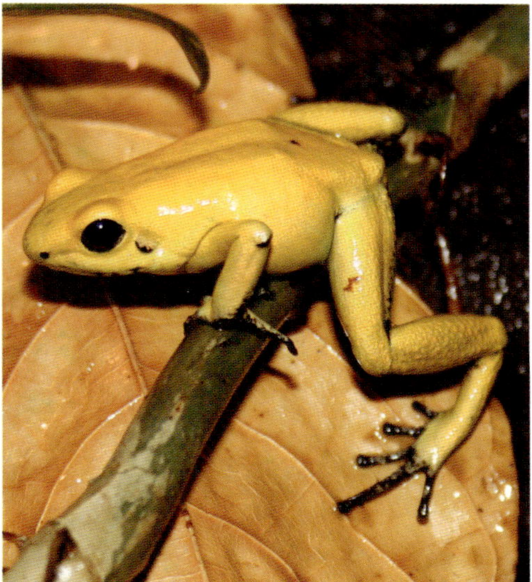

ALL PHOTOGRAPHS:

Golden dart frog

This frog is not venomous (it does not inject poison into its prey), but instead has extremely toxic skin. Indigenous tribes soak their hunting arrows in the poison exuded by frogs, which they catch and expose to gentle heat. The poison remains potent for two years or more.

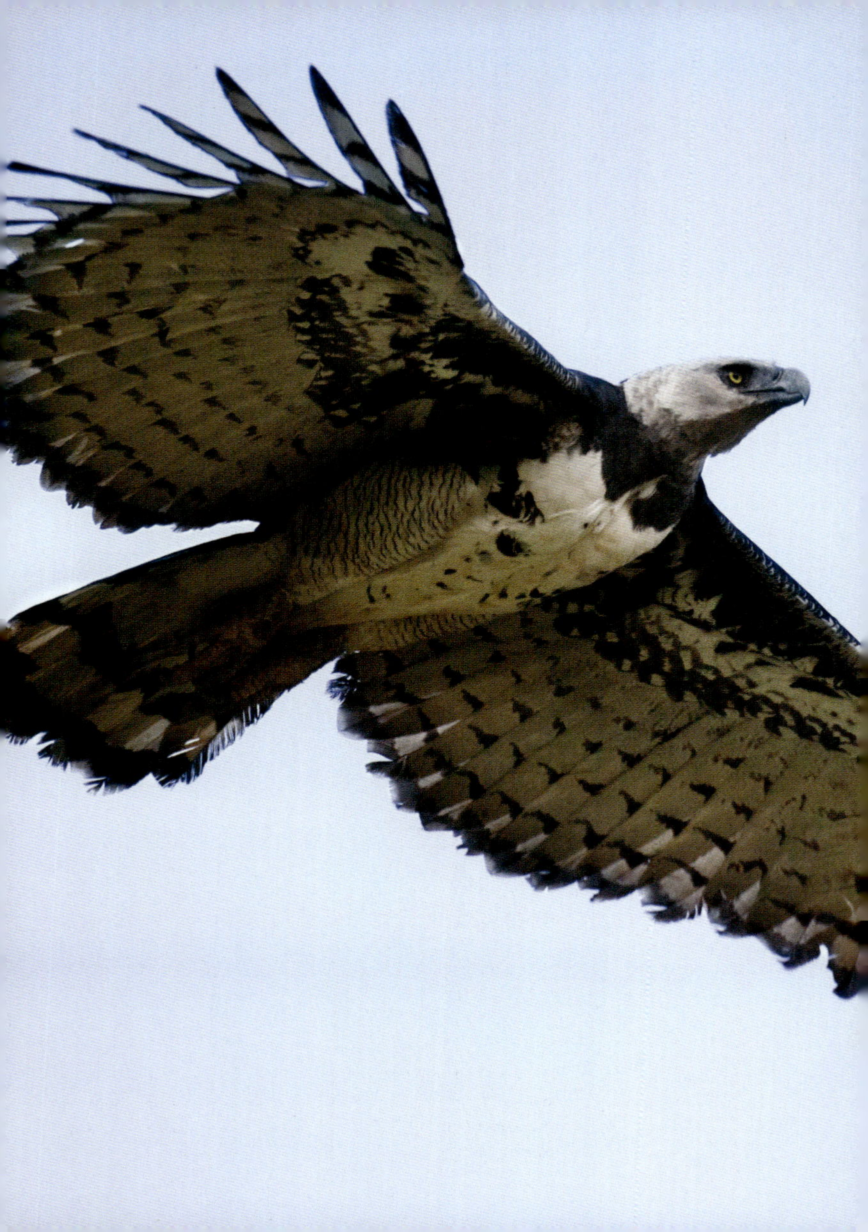

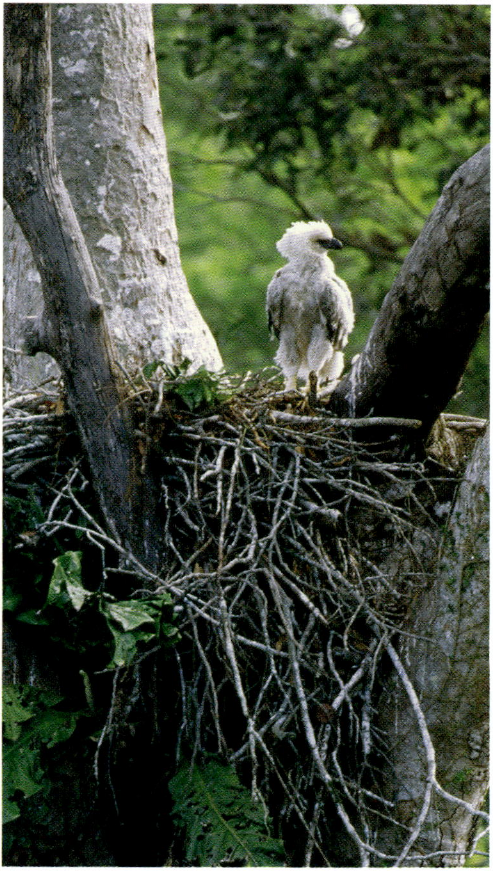

ALL PHOTOGRAPHS:

Harpy eagle

This immensely powerful raptor lives in tropical forests where it hunts prey, such as sloths and monkeys, among the thick canopy. It is one of the largest eagles in the world with a wing span of up to 2.25m (7.4ft), with huge talons measuring up to 12.5cm (5in). It raises one chick every two or three years.

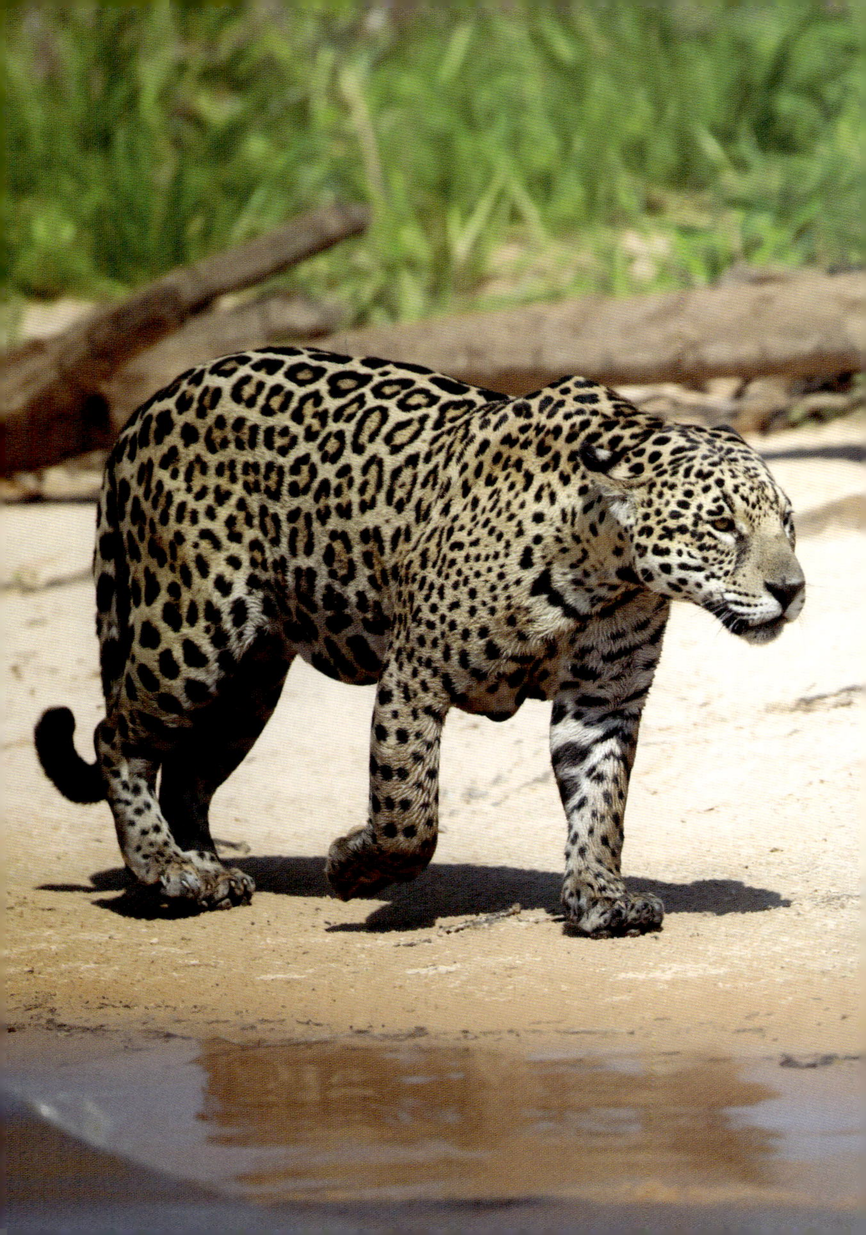

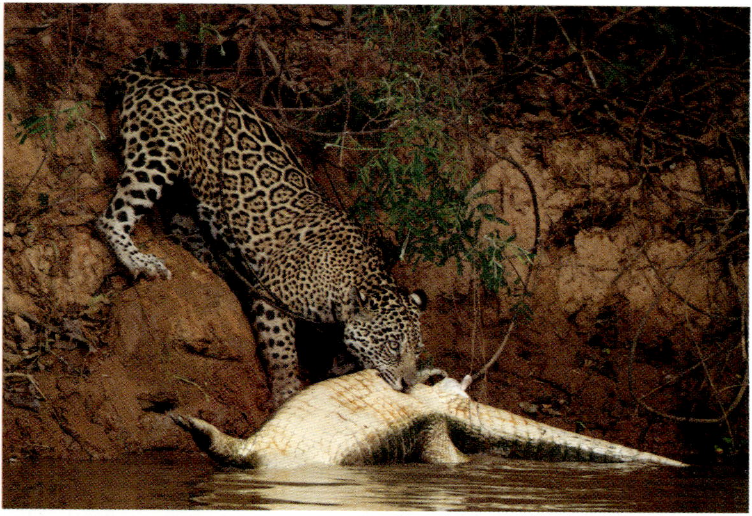

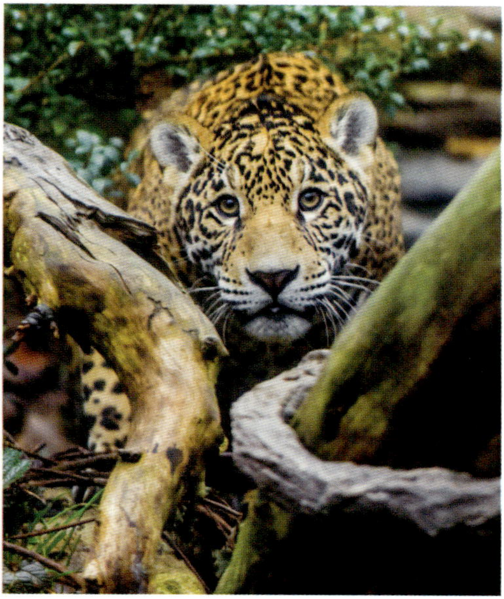

ALL PHOTOGRAPHS:
Jaguar
The jaguar is the largest species of cat in the region, measuring up to 1.8m (6ft). It is a powerful predator, stealthily stalking its prey before the final pounce. Its strong jaws allow it to take prey, such as caiman, which it will catch in water and carry to dry land to consume.

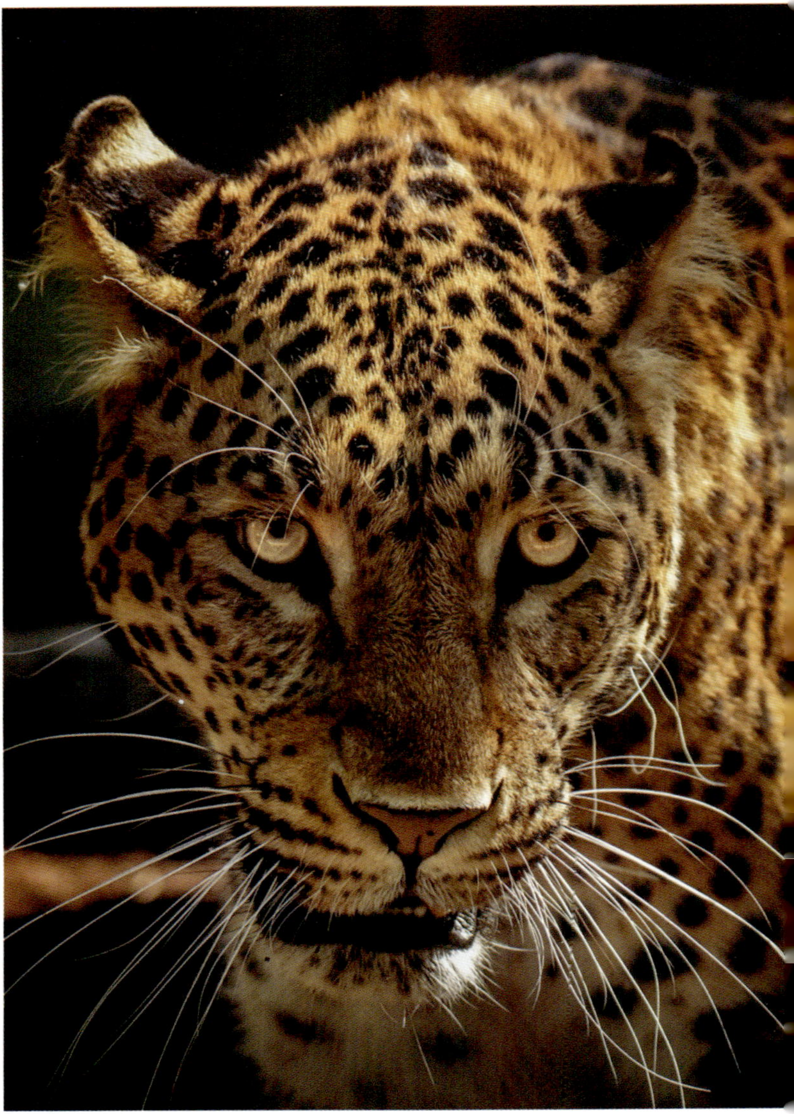

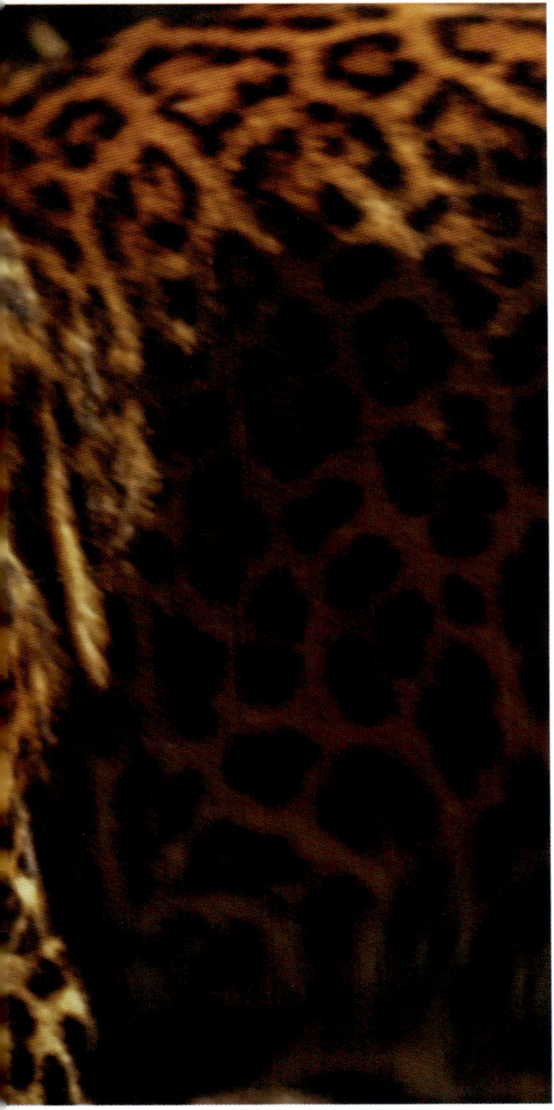

Jaguar

Jaguars have a bite of tremendous force that they use to crush the skulls of mammals or the shells of turtles. They are illegally poached for their body parts, and persecuted by livestock farmers, although habitat loss is by far the greatest threat to their survival. They do not pose a threat to humans.

Maned wolf

So-called in reference to the black mane running from its head and along the back, this long-legged, reddish canine is the largest in the area, measuring 1m (3.3ft). It lives in semi-open grasslands across central South America, where it hunts small mammals and birds, and forages for fruit and tubers.

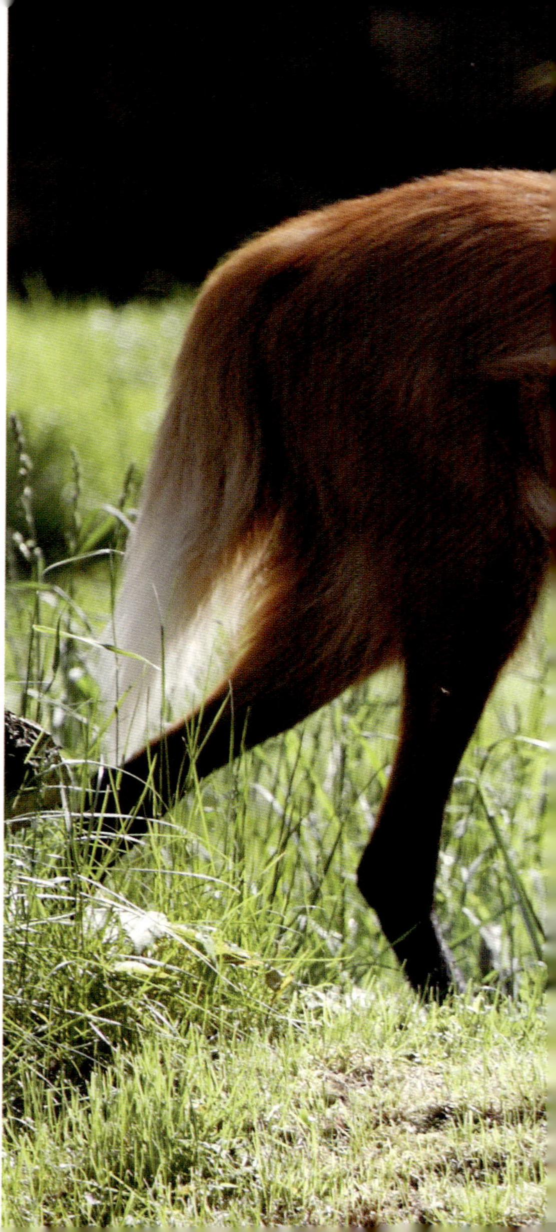

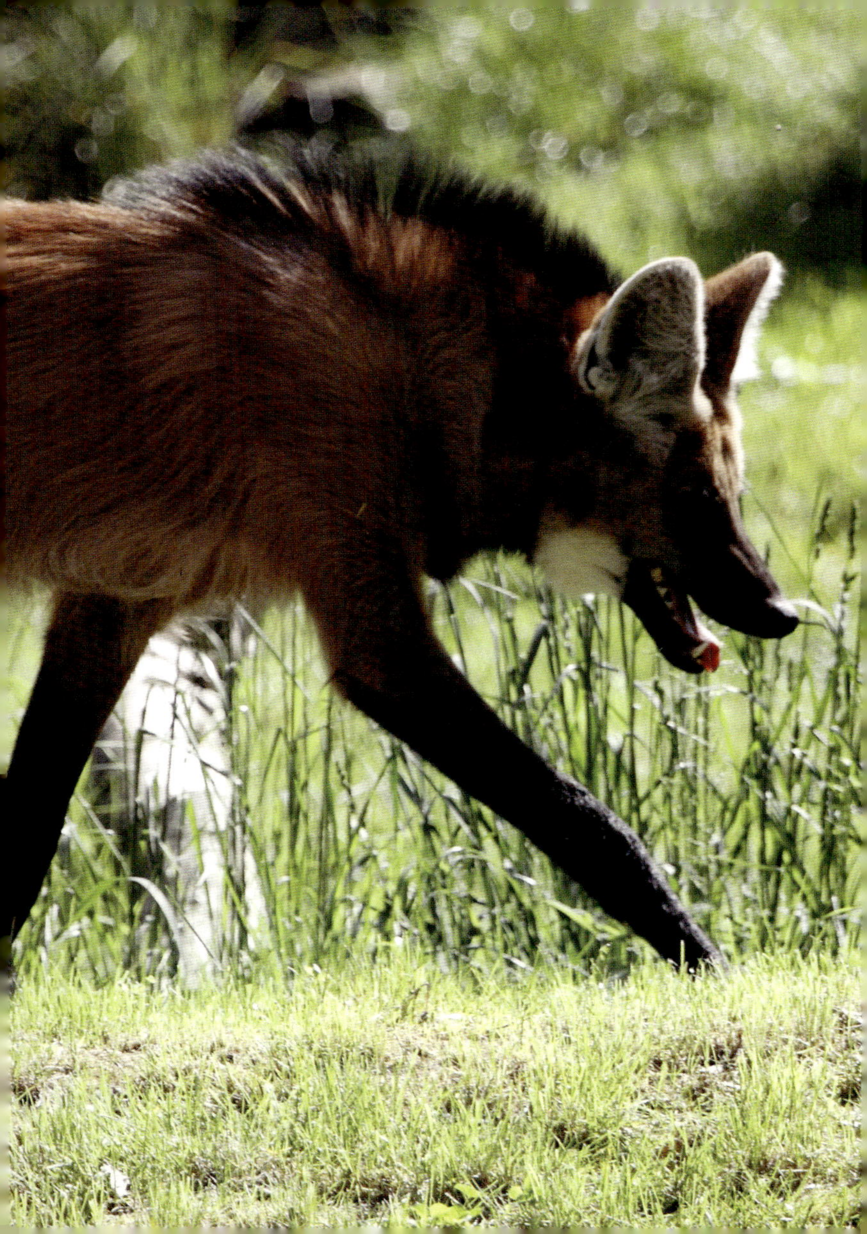

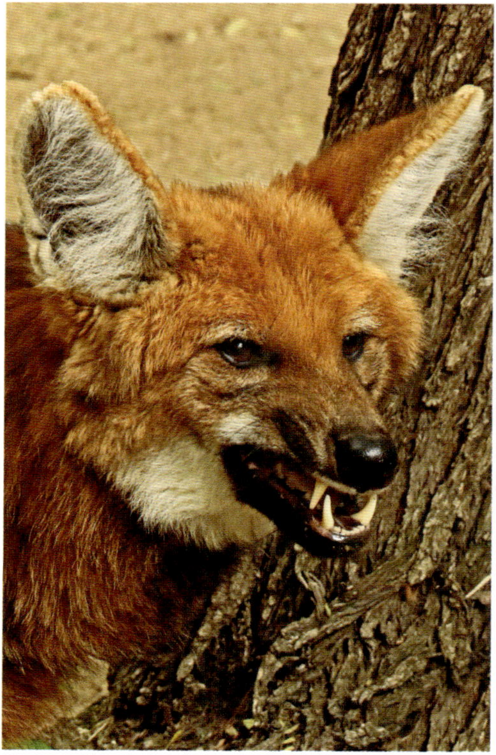

ALL PHOTOGRAPHS:

Maned wolf

This species is the only surviving member of a distinct evolutionary branch, and is unrelated to any other living canine. It is a solitary and shy animal, avoiding contact with humans. Habitat fragmentation, and conflicts with feral dogs, either by competition or disease, are the main threats to the species.

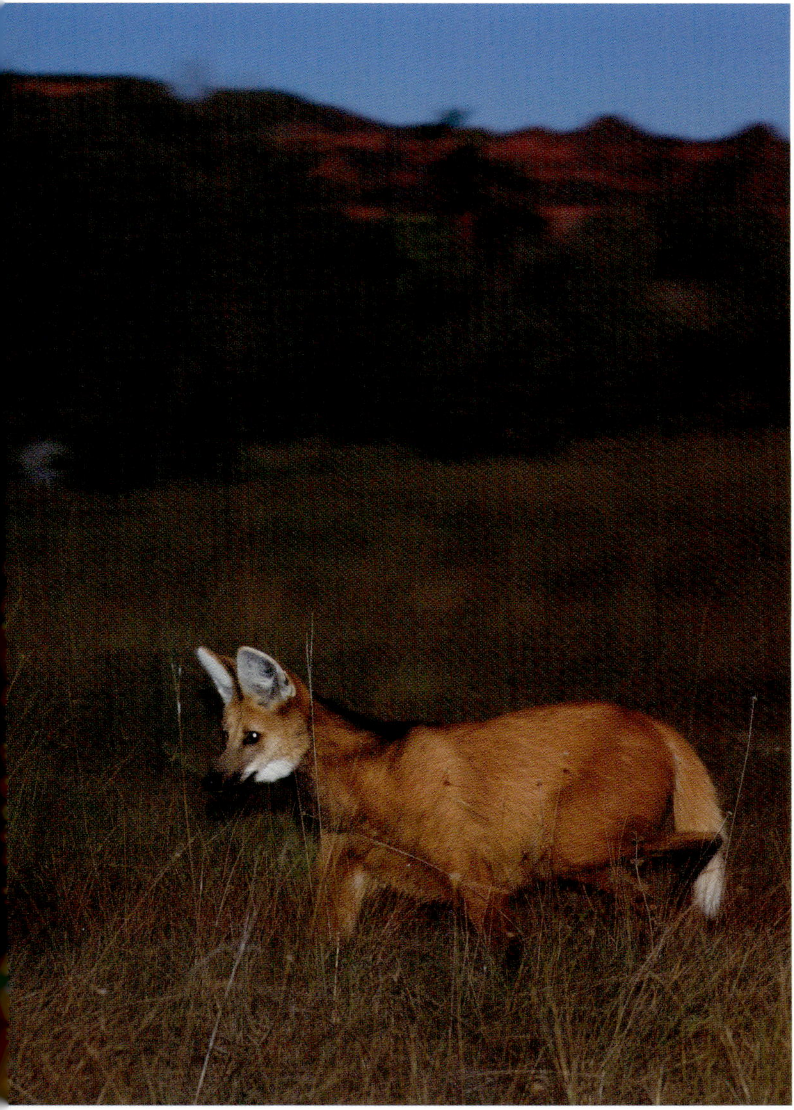

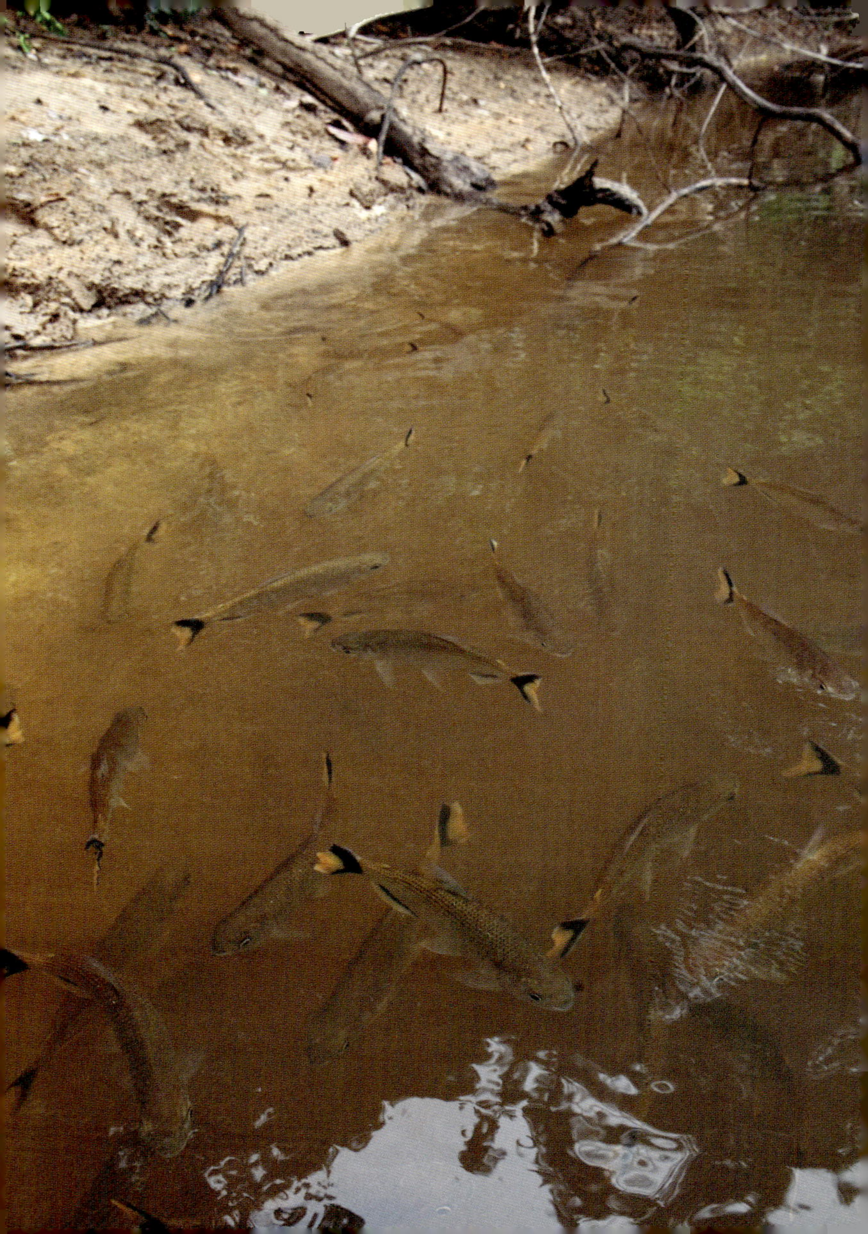

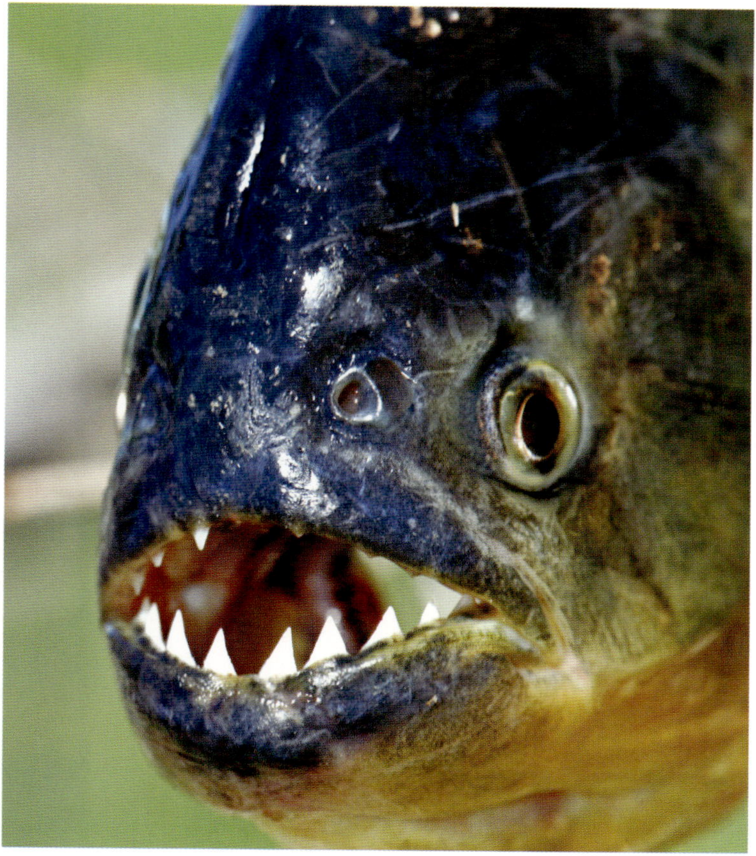

ALL PHOTOGRAPHS:

Piranha

This family of freshwater fish have a fearsome reputation, represented in legend, film and folklore as being a serious predator of humankind. In fact, these fish are largely harmless to humans, feeding instead on smaller fish and plant matter. Their row of triangular sharp teeth is used to shear flesh from their prey.

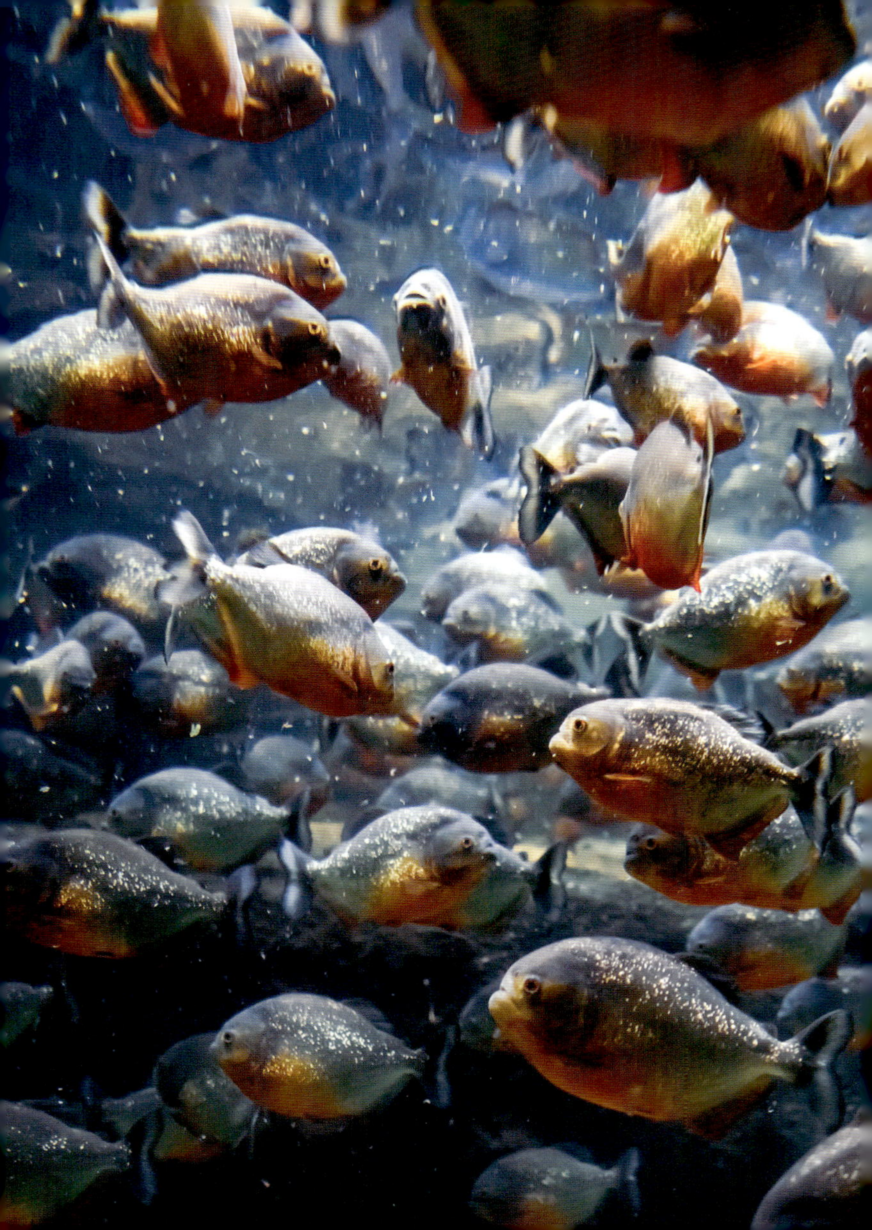

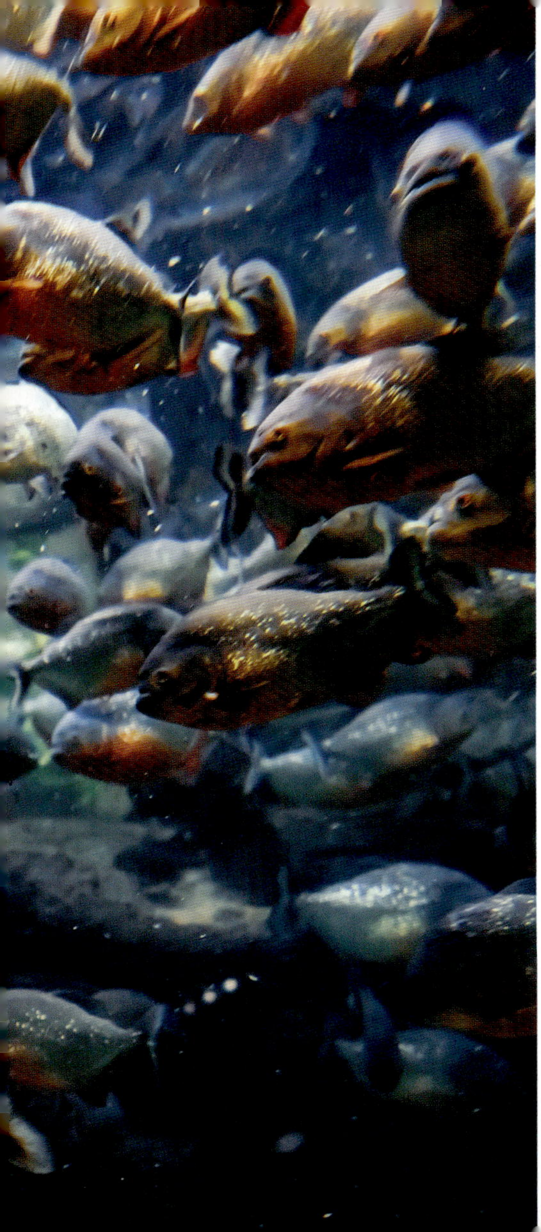

School of red-bellied piranhas

Piranhas form schools as a protective strategy against predators, such as caimans and some water birds. Some piranha species grow to over 30cm (12in) and, rather than eating humans, are instead eaten by them, as a popular culinary dish.

Strawberry poison-arrow frog

An extremely variable species with several colour morphs, this frog lives in isolated populations. The most common colour variation is uniform bright red and 'blue jeans', shown here, which gives the species its name. This frog is small, measuring only 20mm (0.8in), and can be found in humid habitats of Central America.

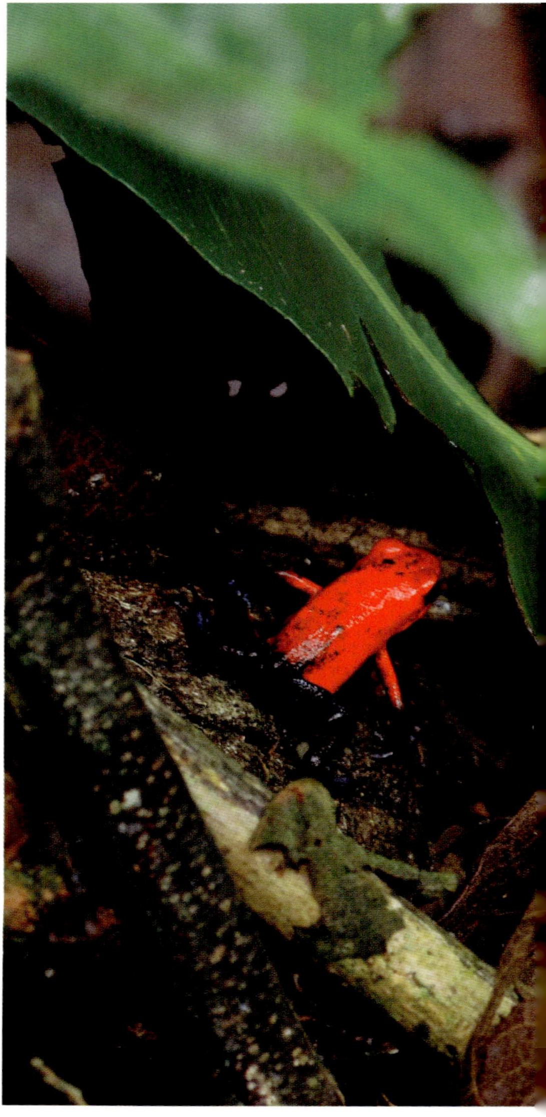

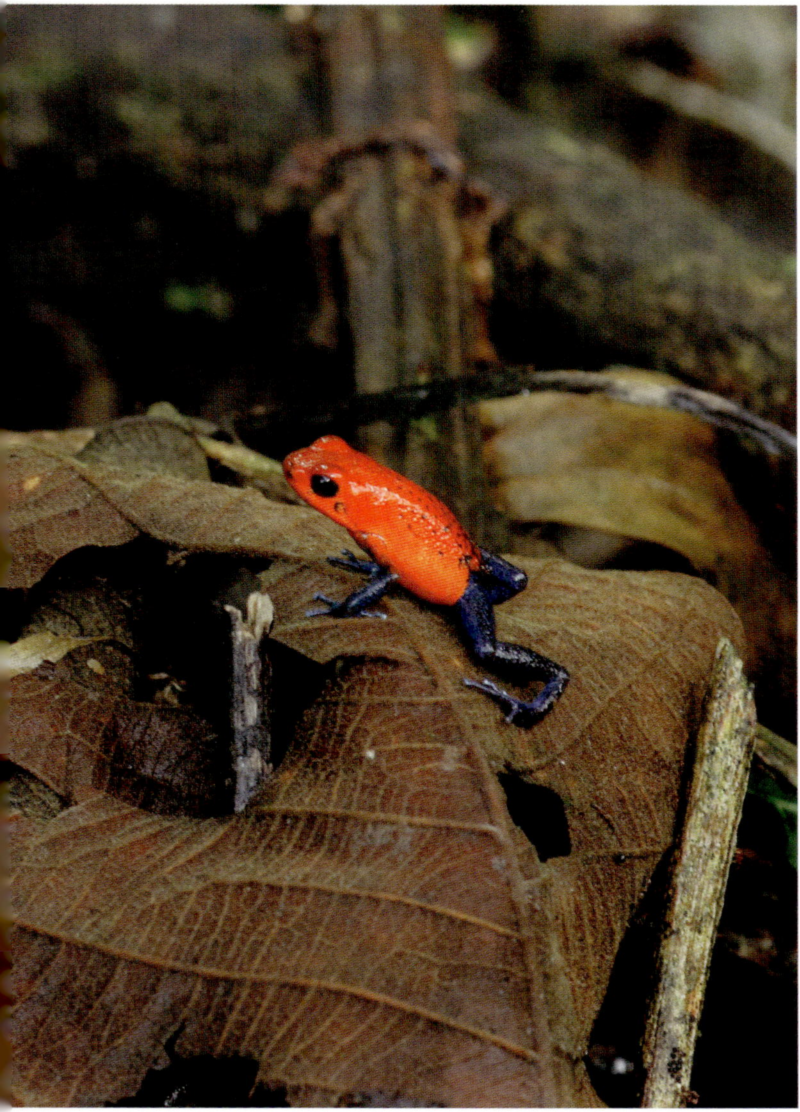

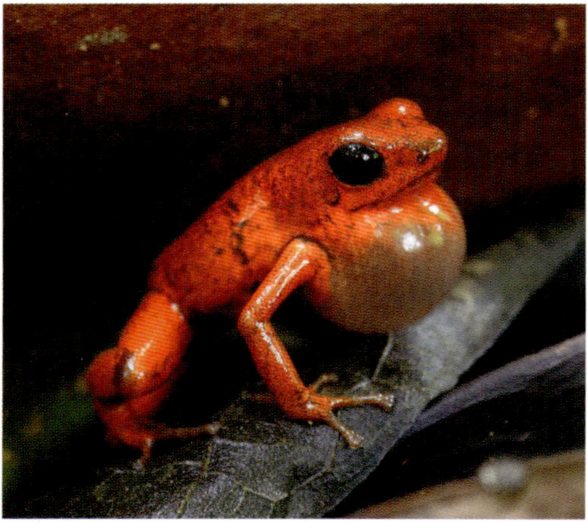

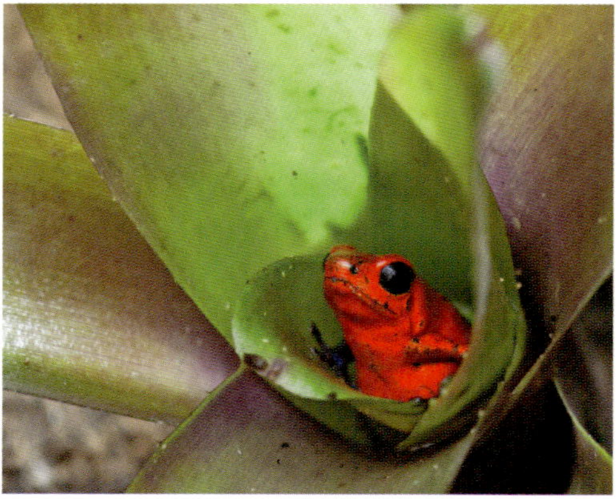

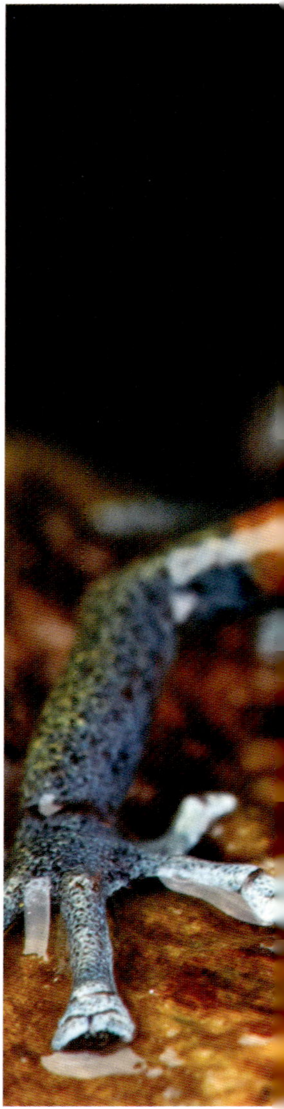

ALL PHOTOGRAPHS:

Strawberry poison-arrow frog

Adult toxicity is obtained from its diet of ants and mites, with the bright colouration serving as a warning to predators. An essential element in the development of tadpoles involves the female feeding them her own unfertilised eggs. In this way, the young obtain sufficient toxin to prevent predation.

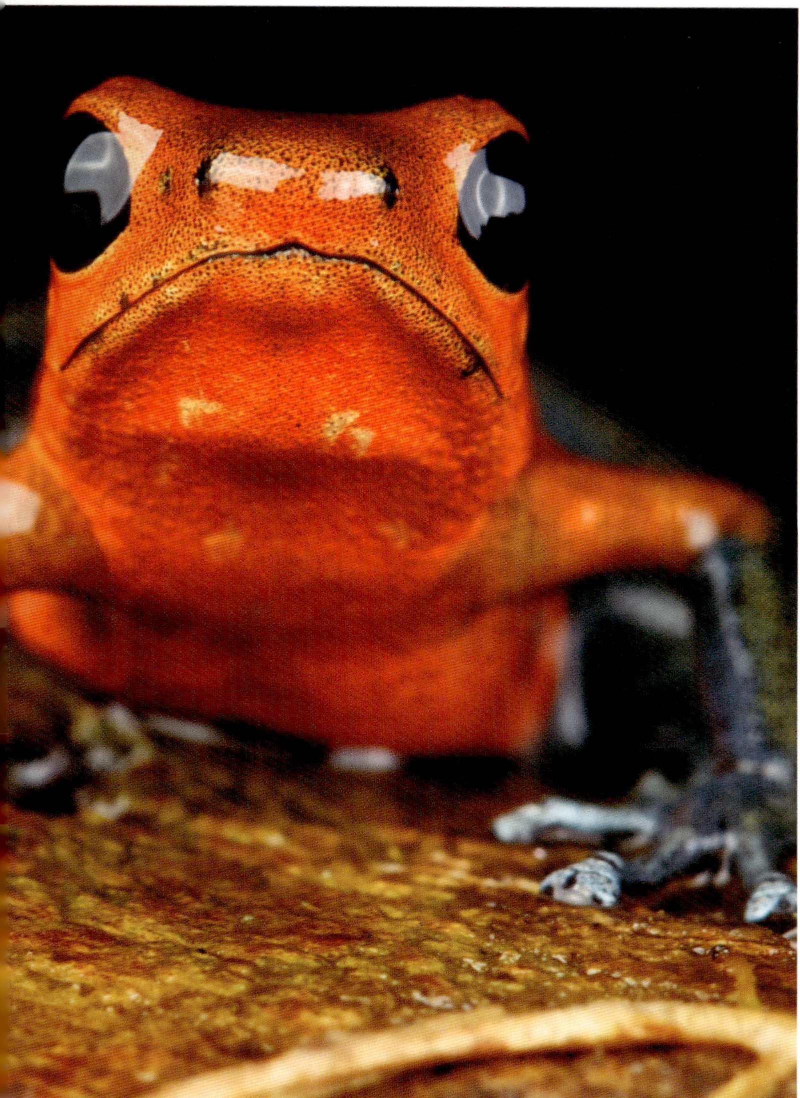

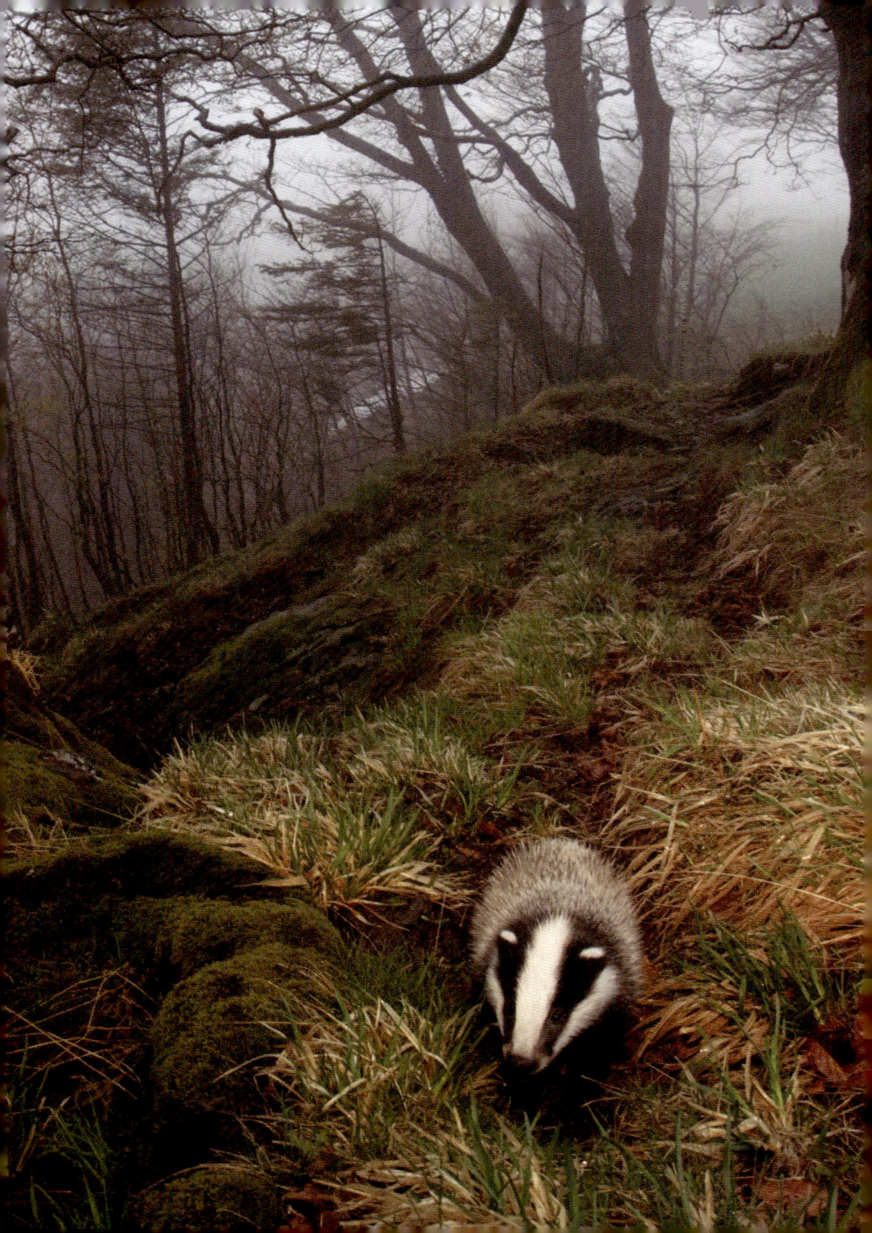

Europe

Europe forms a land mass encompassing mainly temperate regions. It is the second-smallest continent, but has the highest population density, spread over 50 countries. Among these are the smallest recognised state, the Vatican City, and the largest country in the world, Russia.

It has a range of diverse habitats reaching to the Arctic in the north, through central moderate zones and down to the Mediterranean Sea in the south. It features several mountainous areas, such as the Alps, Pyrenees and Caucasus, as well as lowland forests, wetlands and large areas of agricultural land.

Increasing population and associated urbanisation results in much pressure on wildlife and wild habitats. In some countries, such as the UK, all apex predators have been exterminated over past centuries, and the remaining wildlife is under constant threat.

To balance this, in some areas of mainland Europe, changes to rural lifestyles and economics, with an associated movement of populations to established cities, has allowed recolonisation by wolves, bears and lynx.

There are ambitious plans to rewild as much of the continent as possible, while minimising conflict between wild animals and humans. Much is to be done, but in this chapter, you will find spectacular images and interesting facts about the most iconic and charismatic of Europe's wild inhabitants.

OPPOSITE:

Badger foraging in woodland

The term 'badger' is thought to refer to the blaze of white (the 'badge') on the forehead of this thick set, powerful animal. It is otherwise covered in coarse, grey hair.

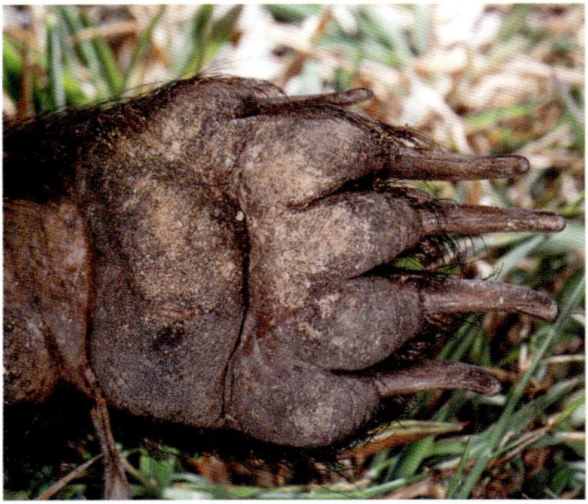

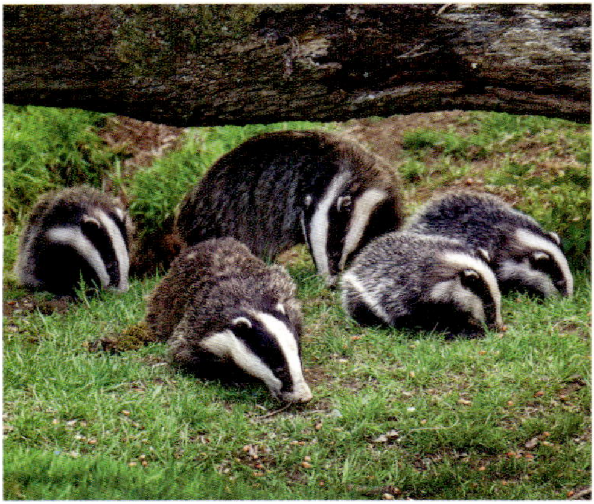

ALL PHOTOGRAPHS:

Badger

Badgers have a maximum body length of 90cm (3ft), and are social animals, living in dens (setts) occupied by several adults and, in season, their cubs. They have long curved claws ideal for digging, and will use these to raid wasp and bee nests where their thick coats provide protection from stings.

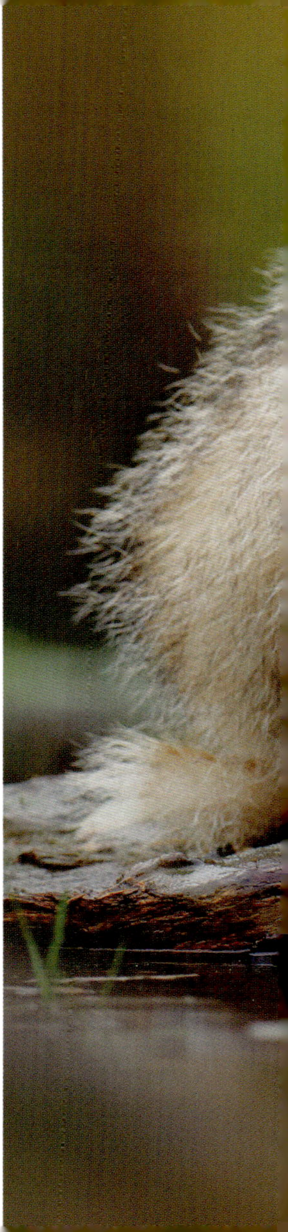

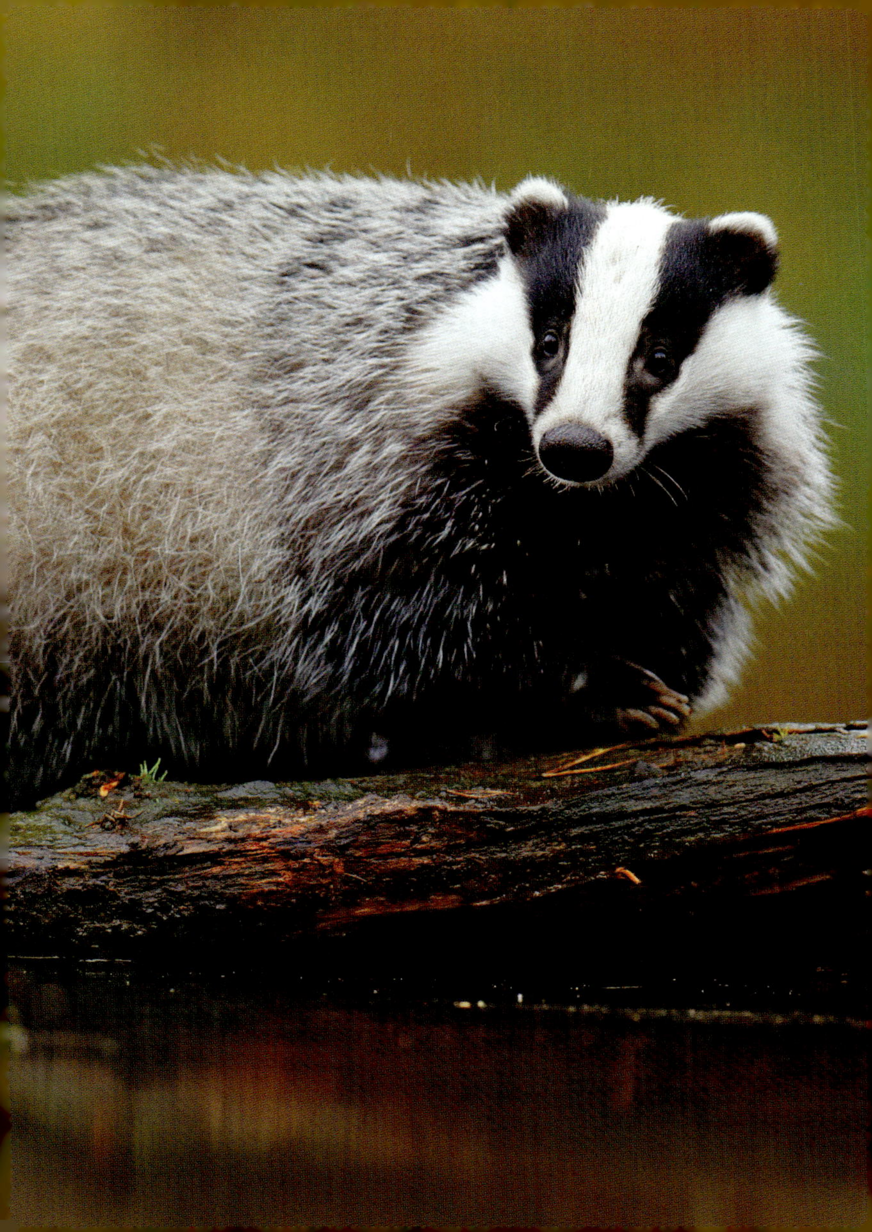

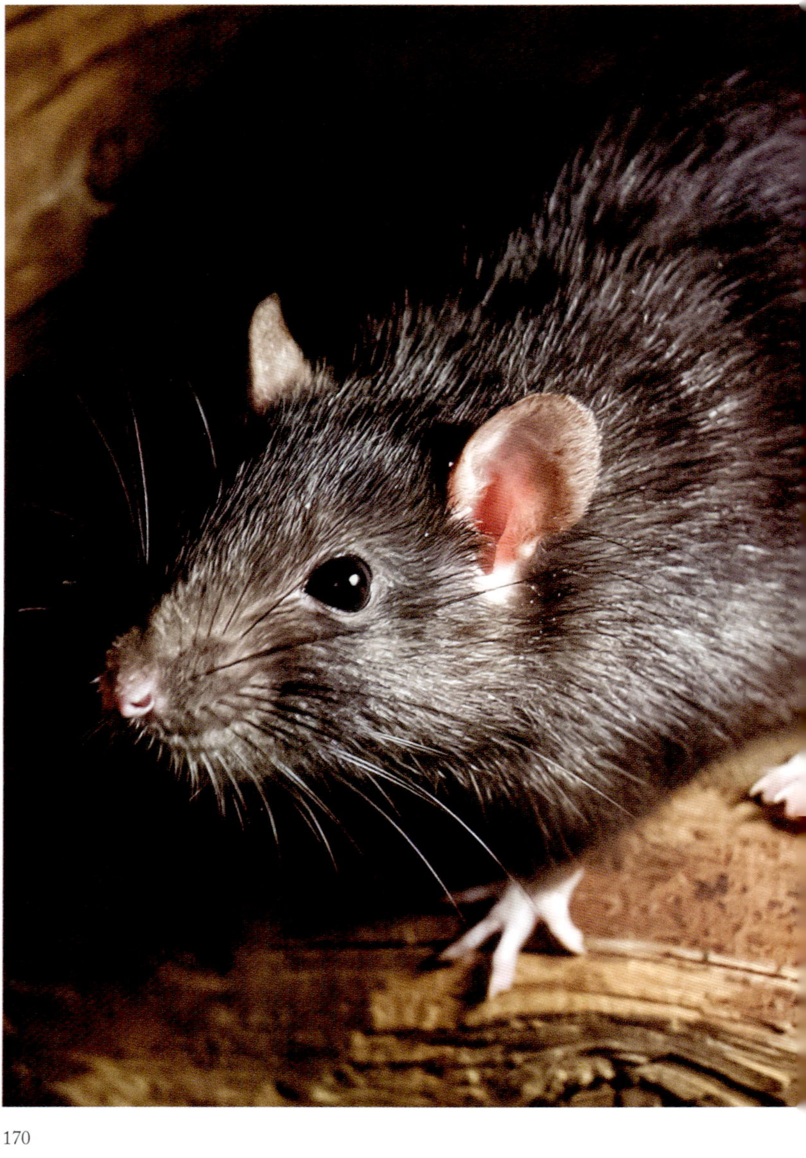

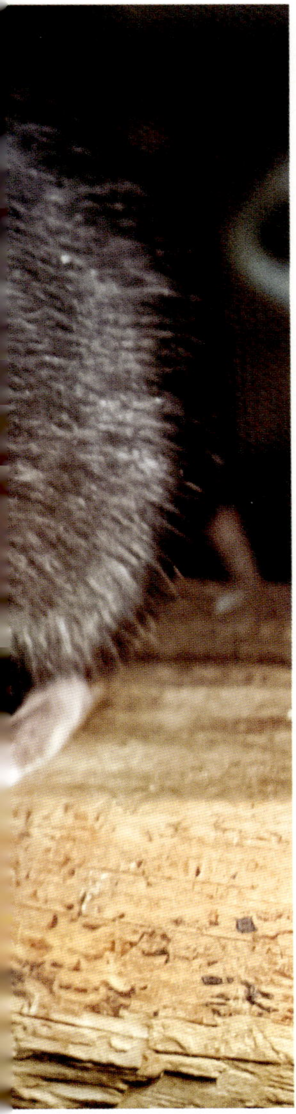

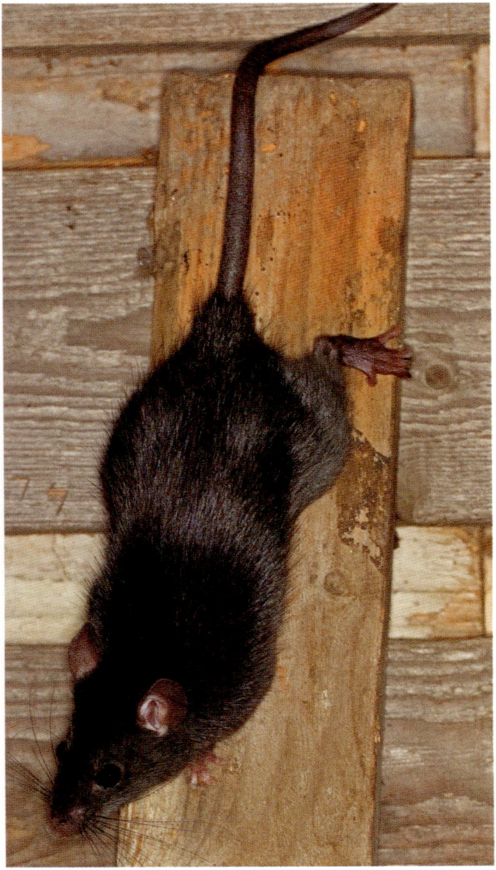

ALL PHOTOGRAPHS:
Black rat
The agile black rat hails originally from Asia, colonising Europe some 2,000 years ago. It has a much slighter build than the more aggressive brown rat, with a more pointed face and hairless ears. Black rat fleas are attributed with spreading the 'Black Death' plague across Europe in the Middle Ages.

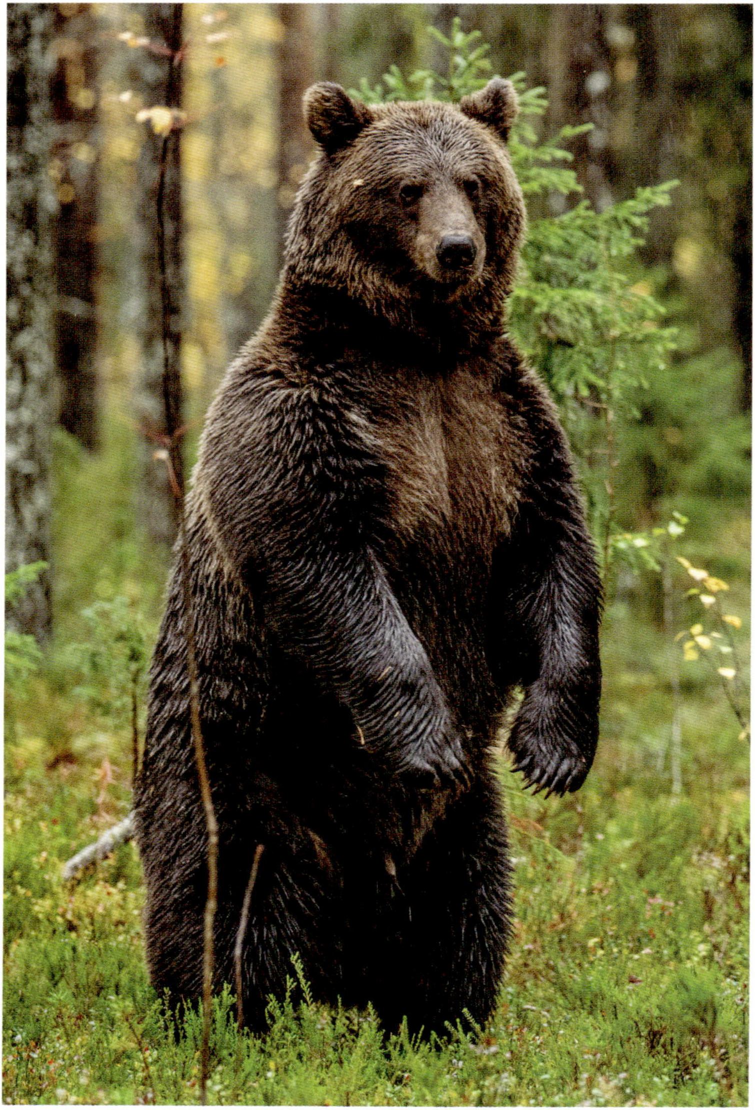

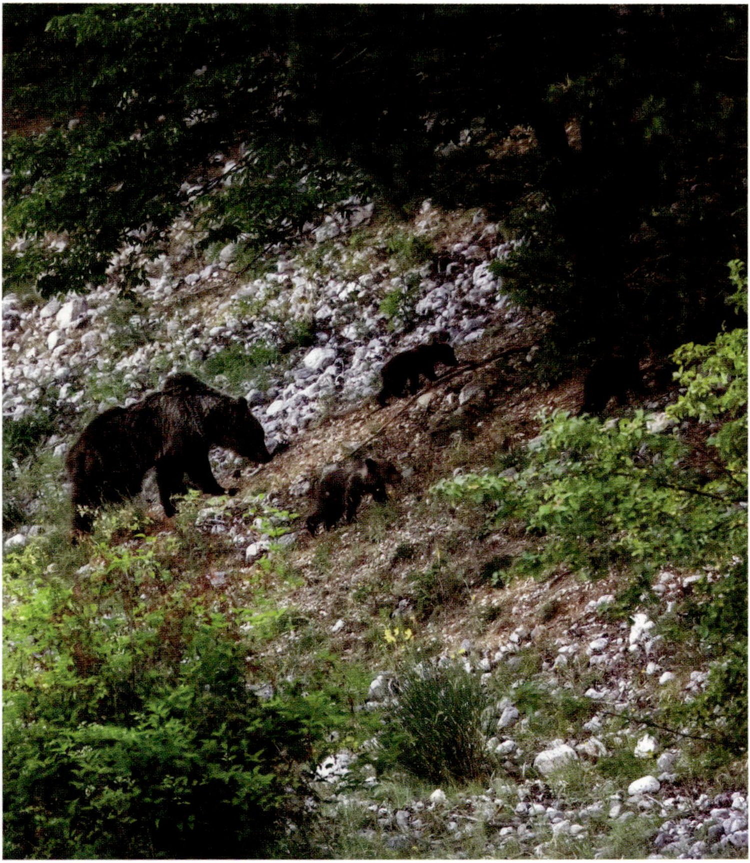

ALL PHOTOGRAPHS:

Brown bear

Covered in uniform brown, reddish or near-black fur, this is a large, omnivorous animal measures up to 2.5m (8.2ft). It has a varied diet of grasses, fruit, nuts, seeds and carrion. Two or three cubs are born between January and March and stay with their mother for up to two years.

Brown bear

This bear mainly inhabits the large, forested areas of Russia, Scandinavia and eastern Europe, with smaller populations elsewhere. Although a fully protected species, in some areas, population expansion is resulting in an increase in encounters with humans, and fatal collisions with vehicles.

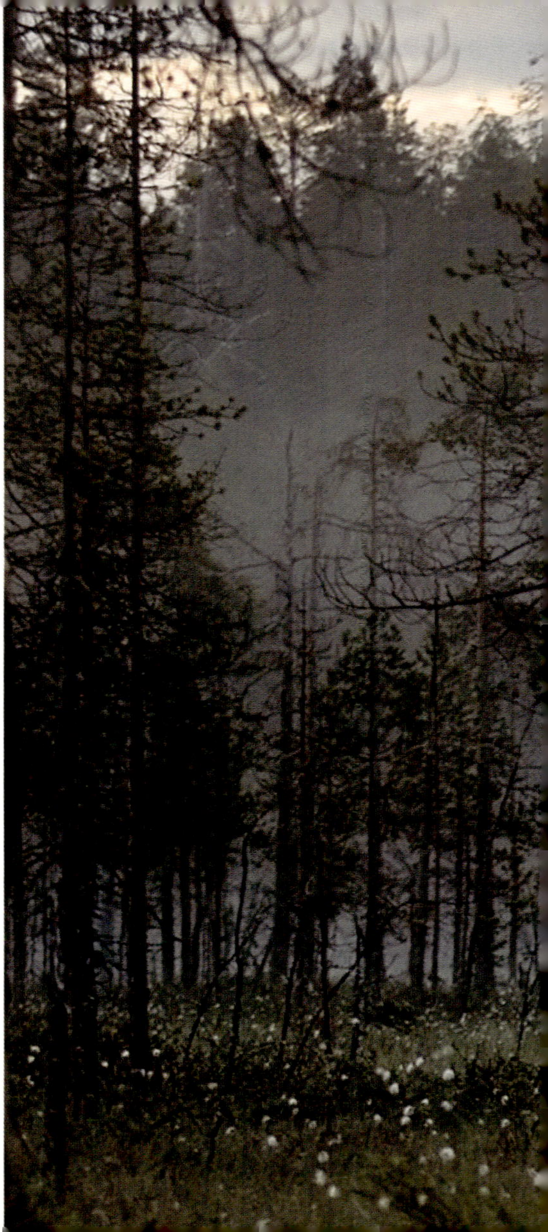

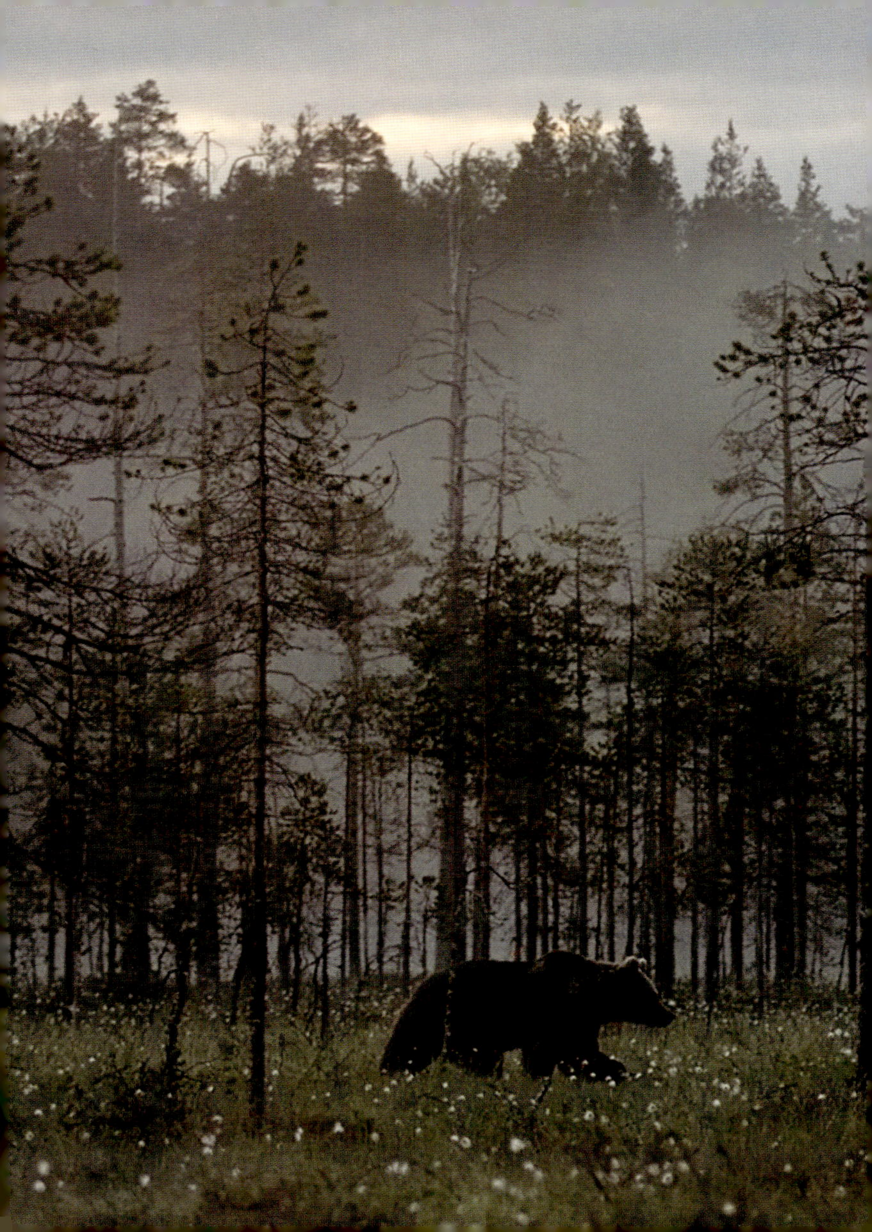

ALL PHOTOGRAPHS:

Eurasian lynx

This shy and elusive animal ranges across Western Europe to Central Asia. It has a brown coat spotted liberally with black, white underparts and a short, black-tipped 'bobbed' tail. The ears have a distinctive tuft of black hair. With a maximum body length of about 1m (3.3ft), it is the largest of the world's four species of lynx.

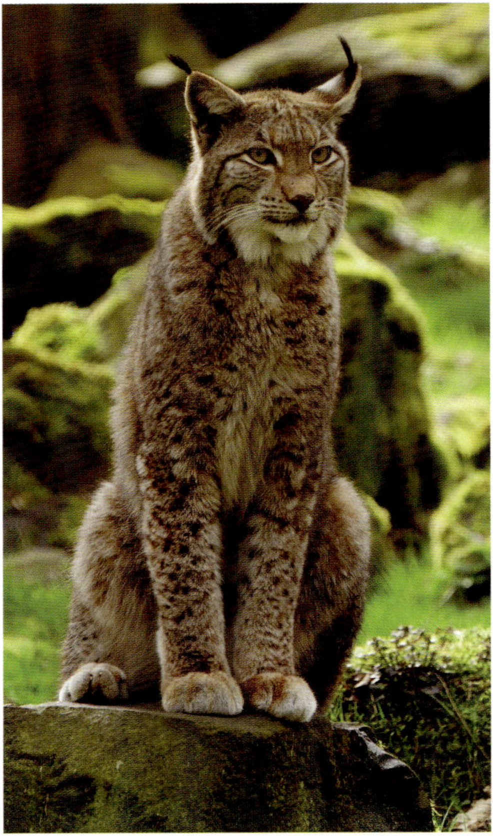

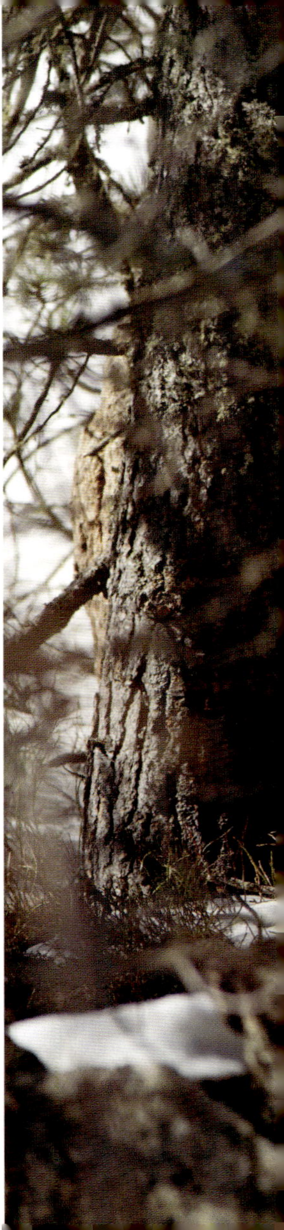

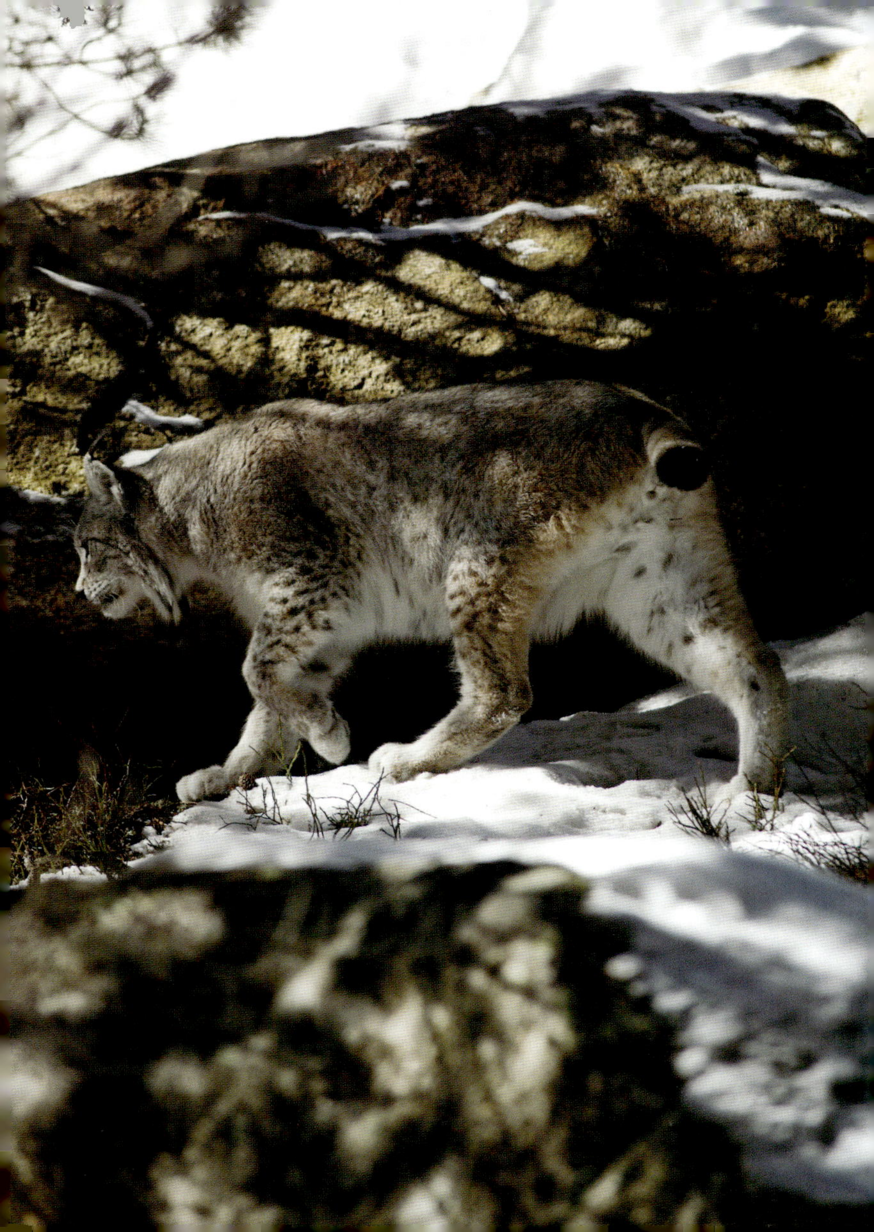

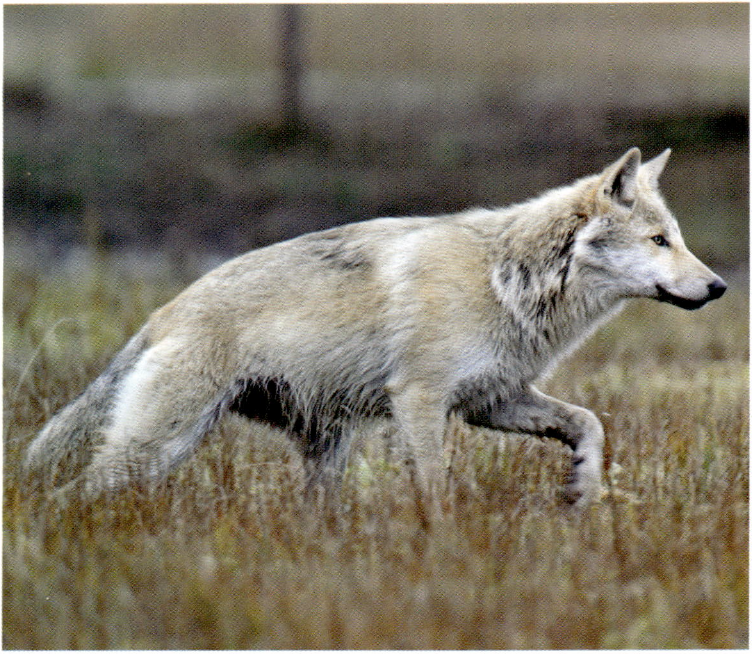

ALL PHOTOGRAPHS:

Grey wolf

Historic persecution dramatically reduced the range and population of this ace predator, which only found refuge in the vast forests of Russia and Eastern Europe. Changes to rural economies, together with conservation activity over the past few decades, has allowed wolves to naturally recolonise many former territories.

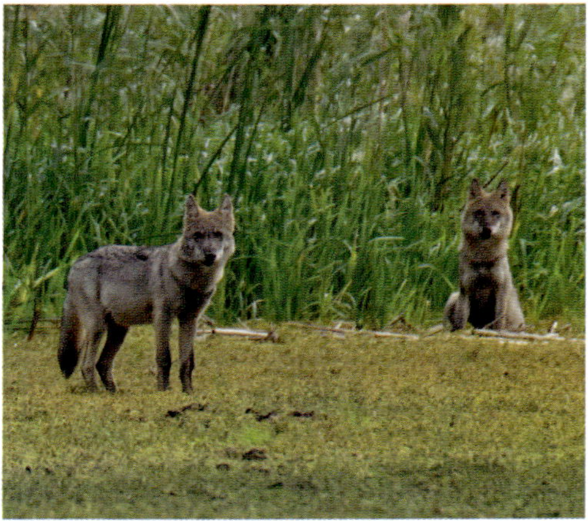

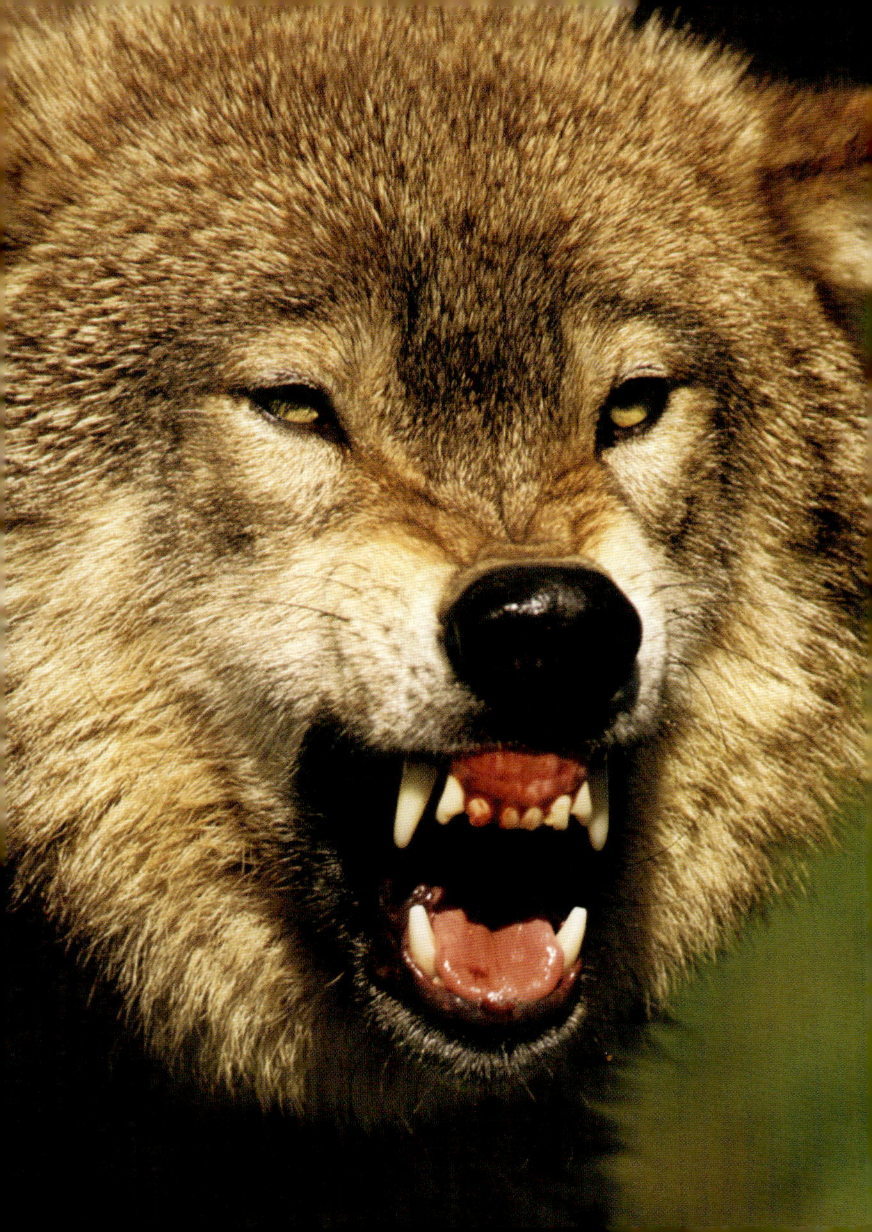

Grey wolf at kill
Wolves are intelligent, social animals living in packs with a hierarchical structure. They hunt deer, goats and wild boar, but will also attack livestock if natural prey is scarce. They feature strongly in mythology, where their howls portent the unleashing of dark forces. But in fact, attacks on humans are extremely rare.

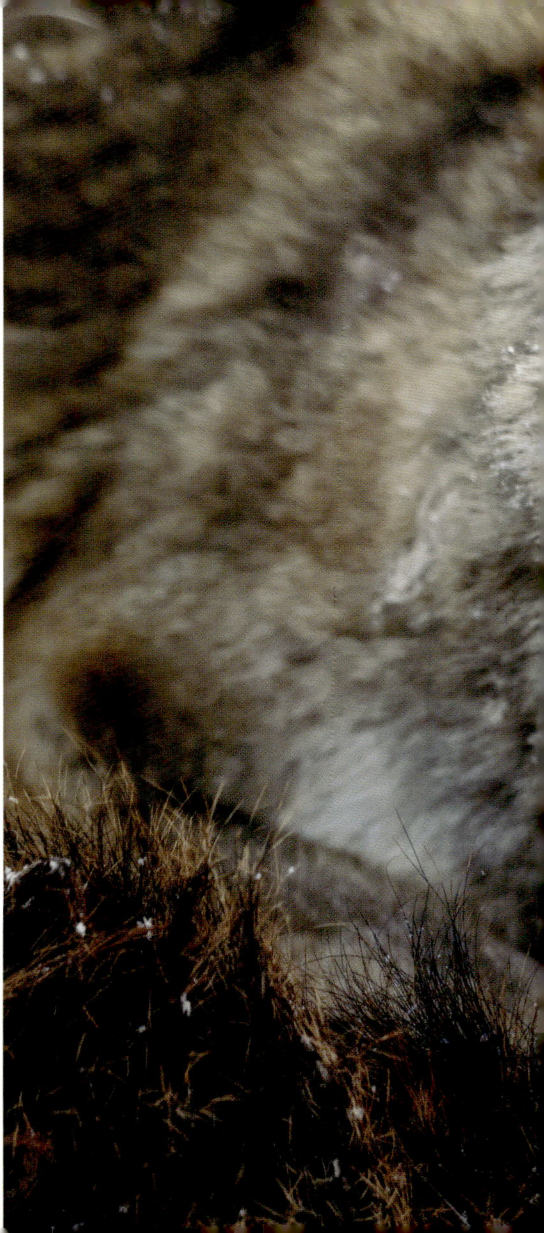

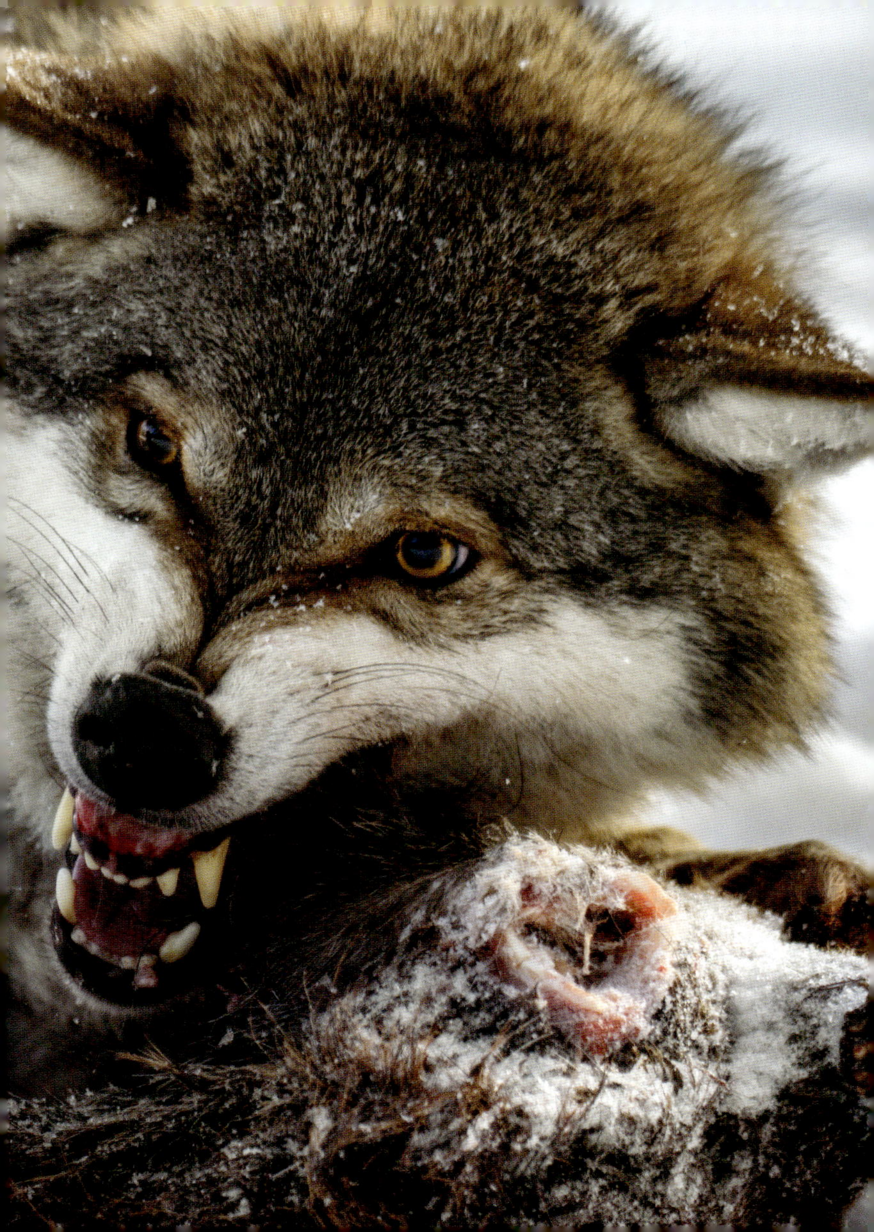

Pike

Pike are long-bodied, predatory freshwater fish, sometimes growing to over 1m (3.3ft) in length. They prey on birds, as well as fish and other aquatic animals, which they ambush by hiding in vegetation, or by slowly manoeuvring to within striking range. Pike are widespread throughout Europe.

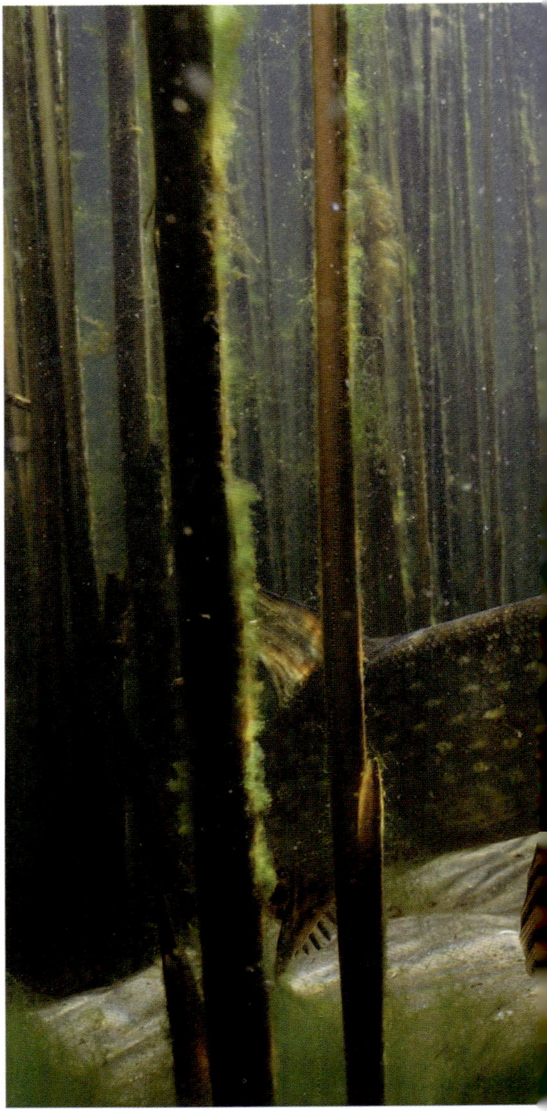

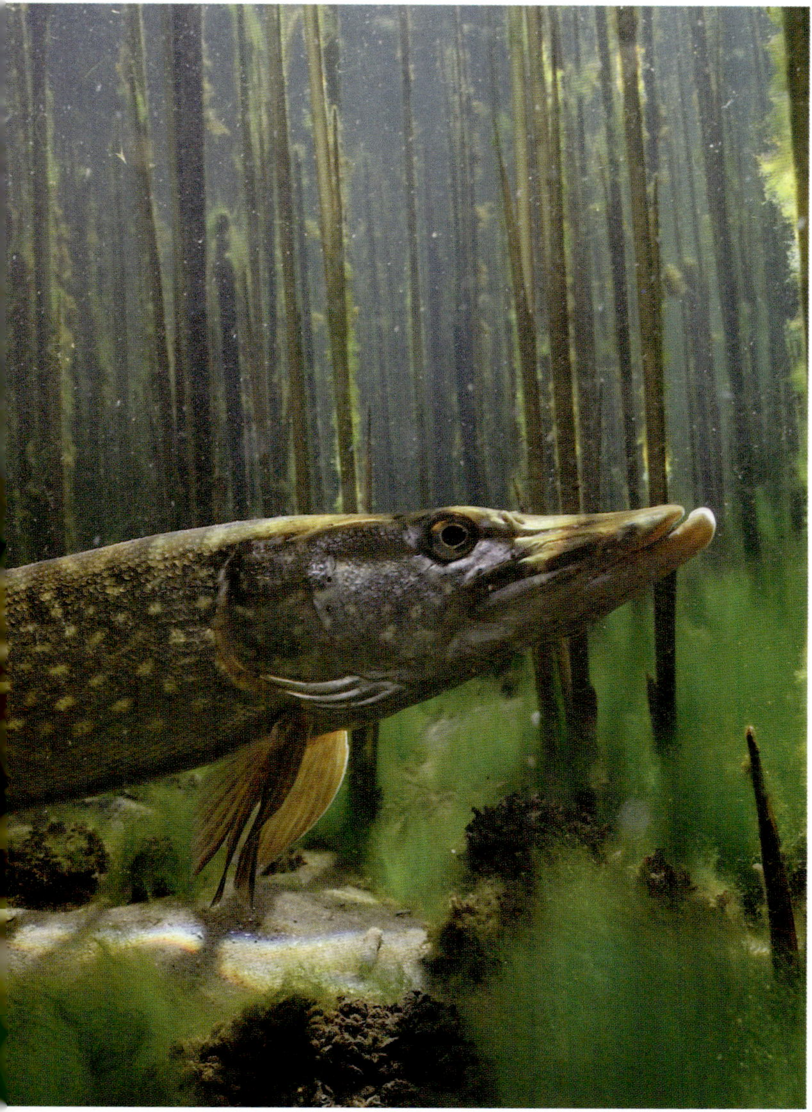

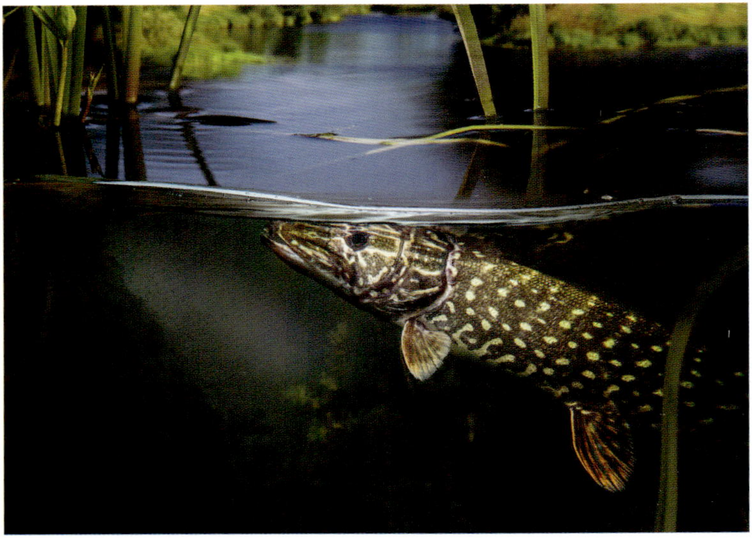

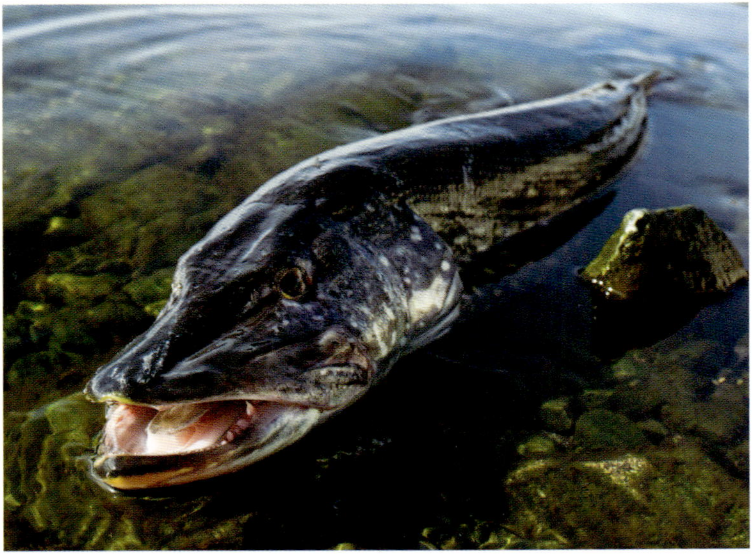

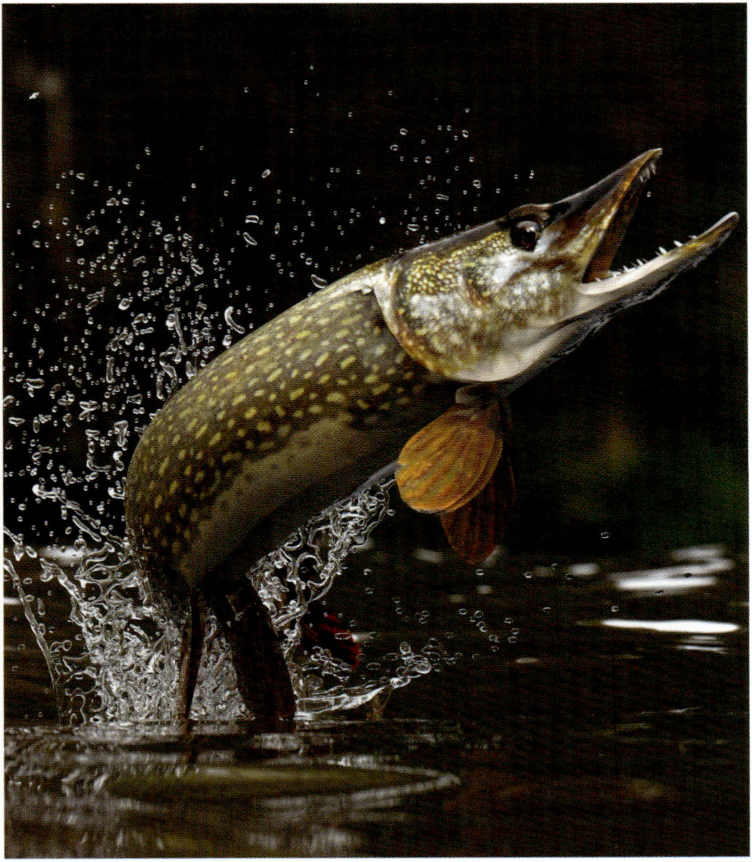

ALL PHOTOGRAPHS:

Pike

Green in colour with a series of cream-coloured spots along the flanks, a pikes' jaws are full of needle-sharp teeth, enabling them to catch and hold onto prey. They will sometimes take quite large animals, such as ducks and small mammals, even leaping from the water to catch their prize.

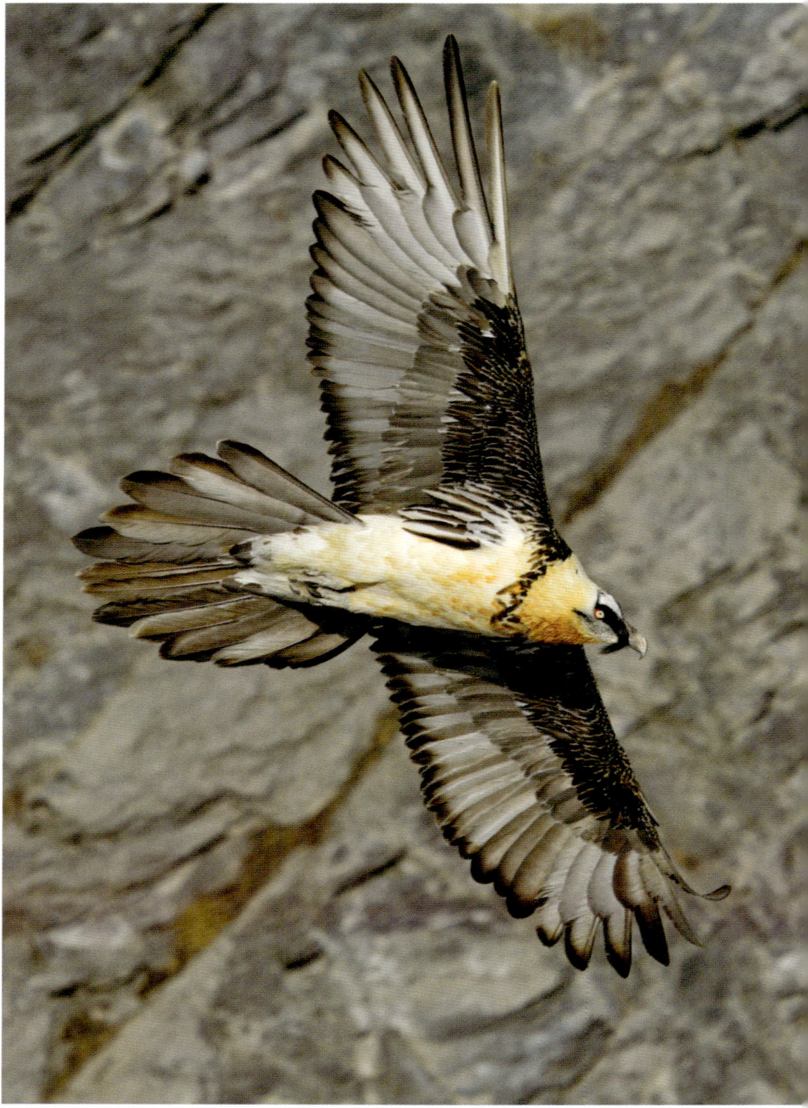

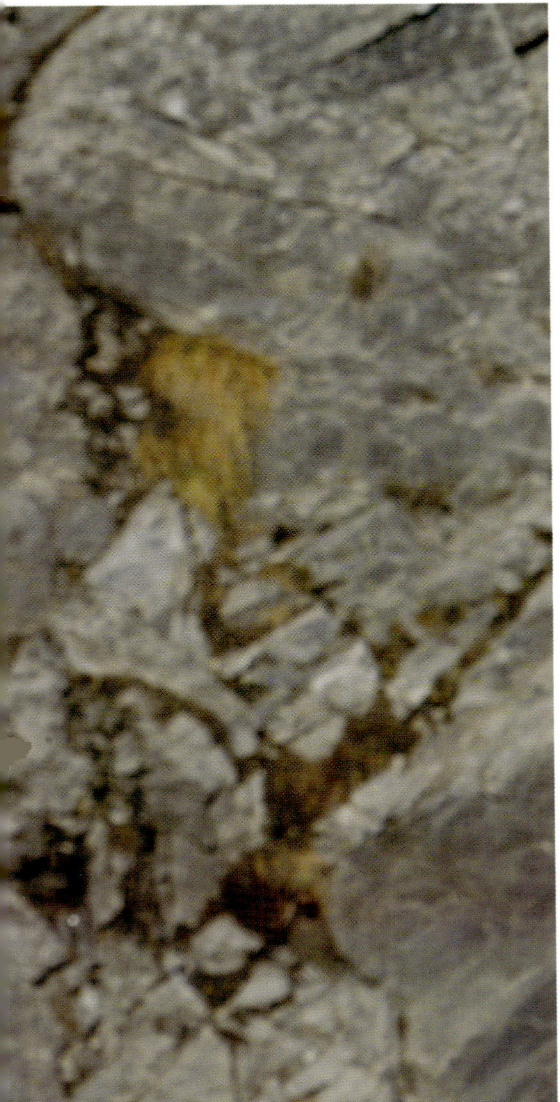

Lammergeier
The lammergeier, or bearded vulture, with a wingspan of over 2.75m (9ft), and a distinctive wedge-shaped tail, inhabits mountainous areas of the continent. It feeds mainly on carrion and bones, and is also known as the 'bone-breaker' due to its habit of dropping bones onto rocks to expose the marrow.

Siberian tiger

This is the largest form of tiger with a very limited European range. It is found only in the far eastern forests of Russia where a population of some 550 animals is thought to exist. Average body length is approximatley 2m (6.5ft), and the pale tawny fur is the thickest of any sub-species.

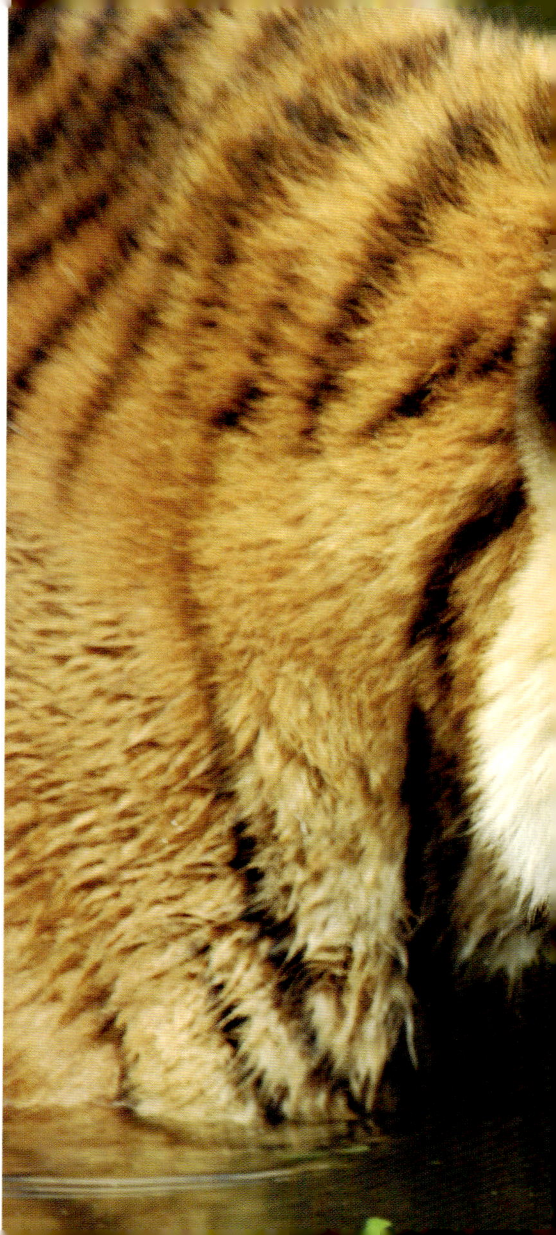

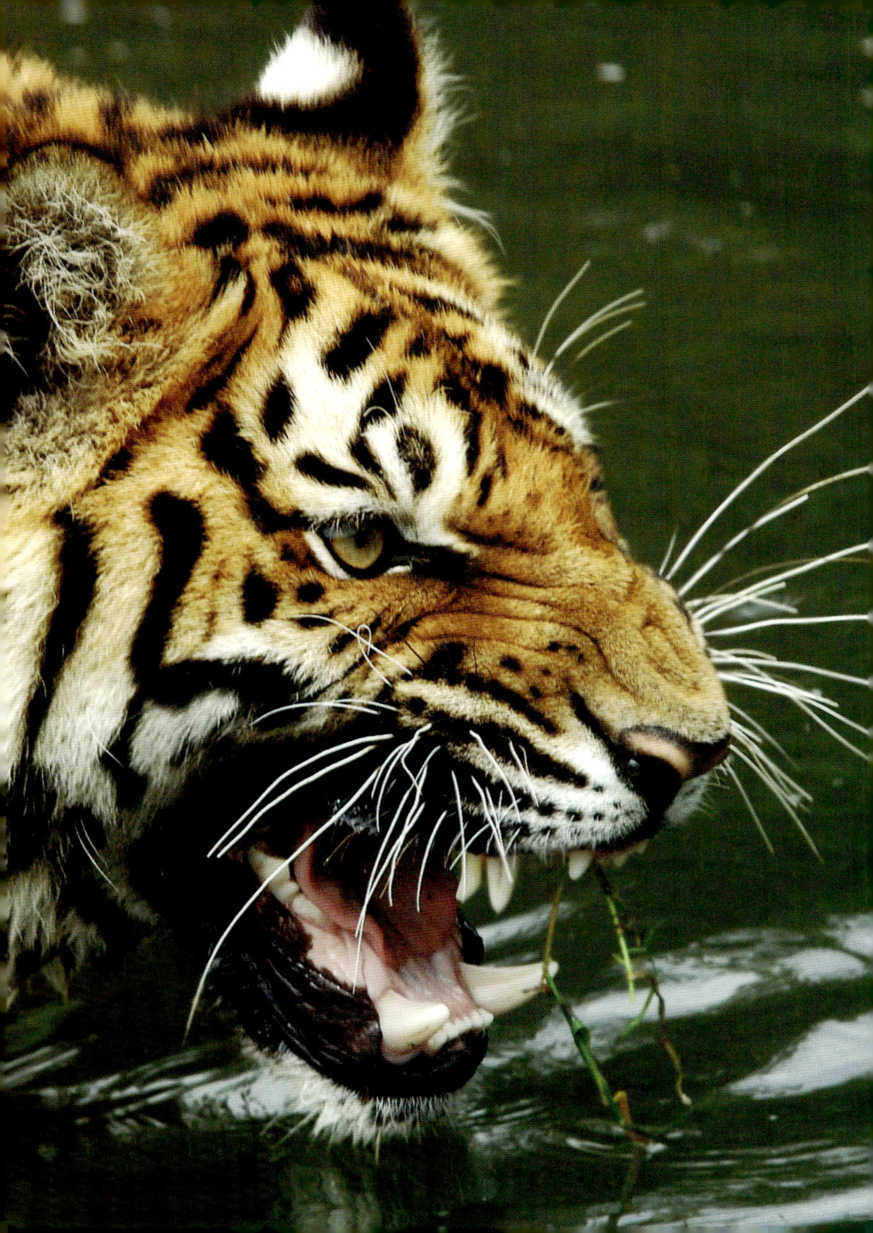

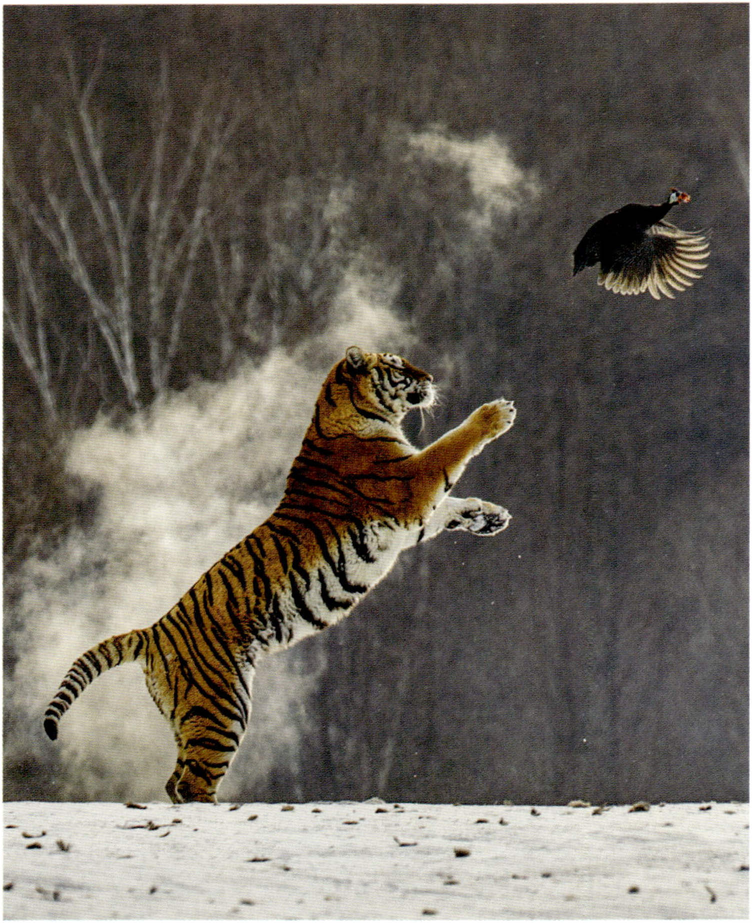

ALL PHOTOGRAPHS:
Siberian tiger
Common prey include deer, moose and wild boar, as well as other small mammals and birds. Competition with brown bears can be intense, with many fatal encounters between them witnessed by scientists, either from direct attack or disputes over food. Conversely, wolf numbers are severely lowered in tiger territories.

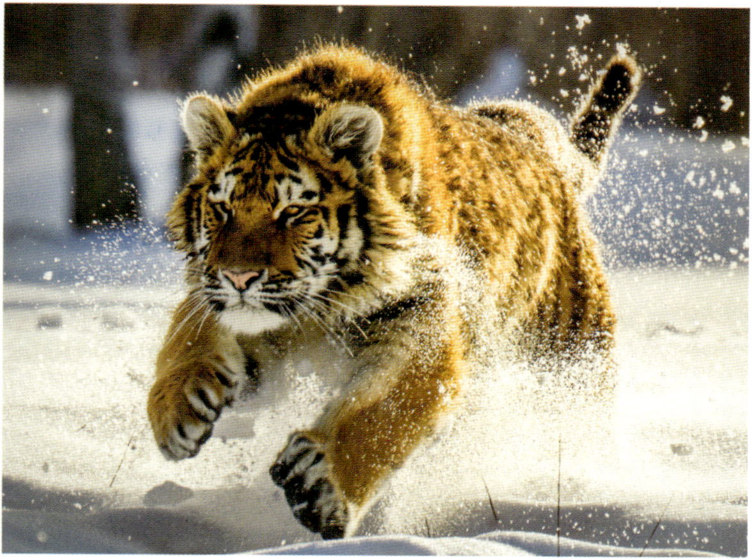

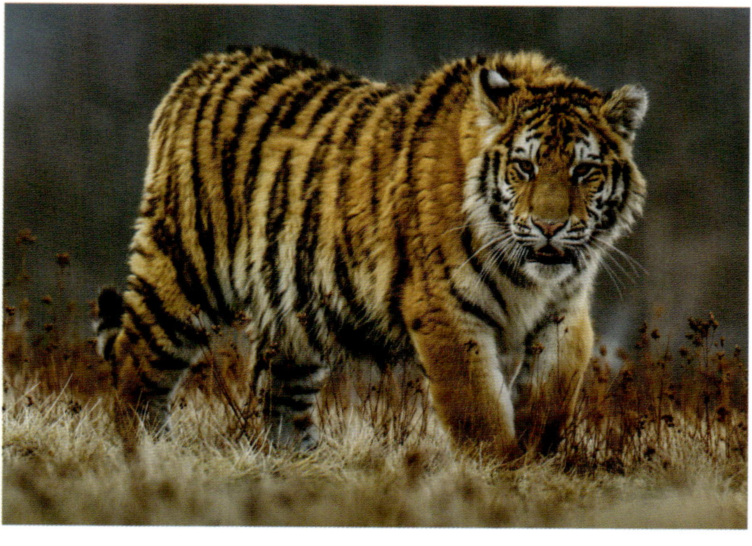

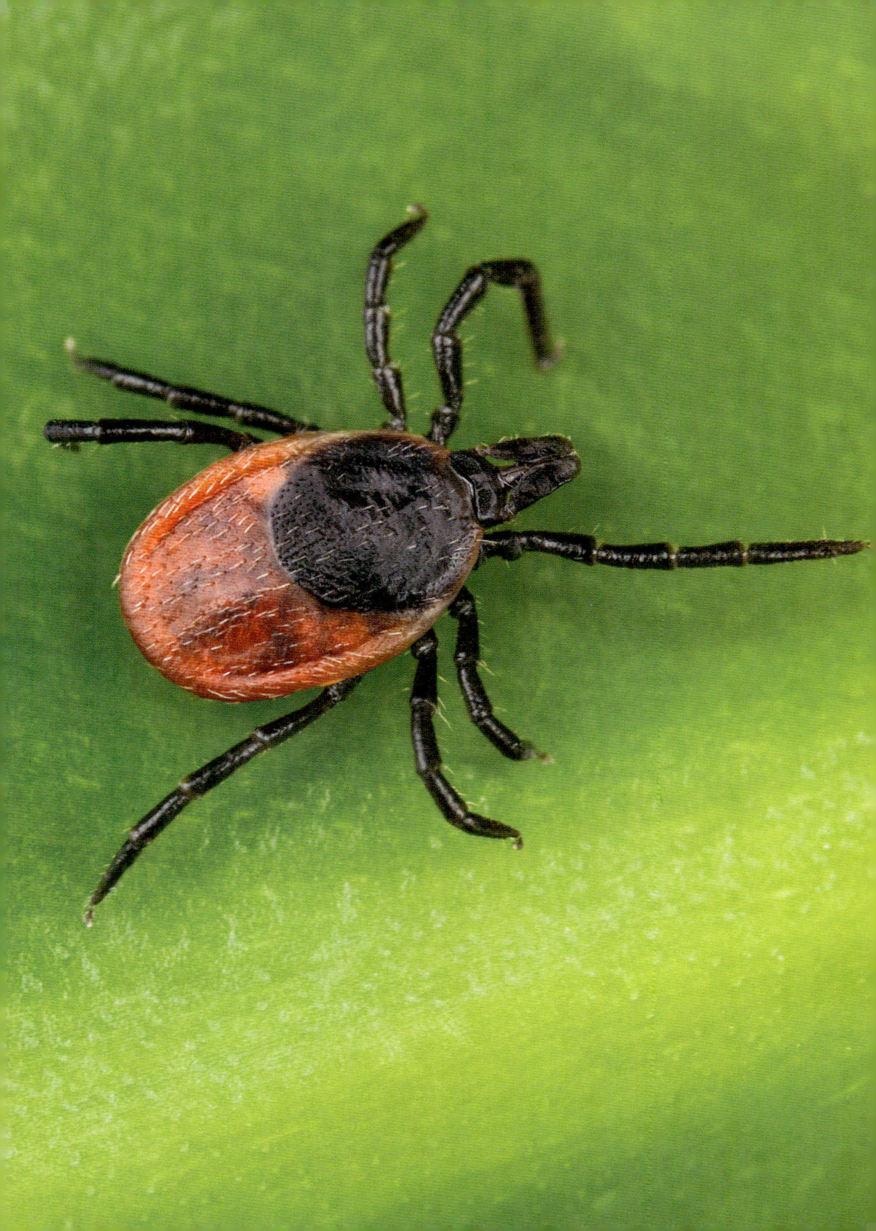

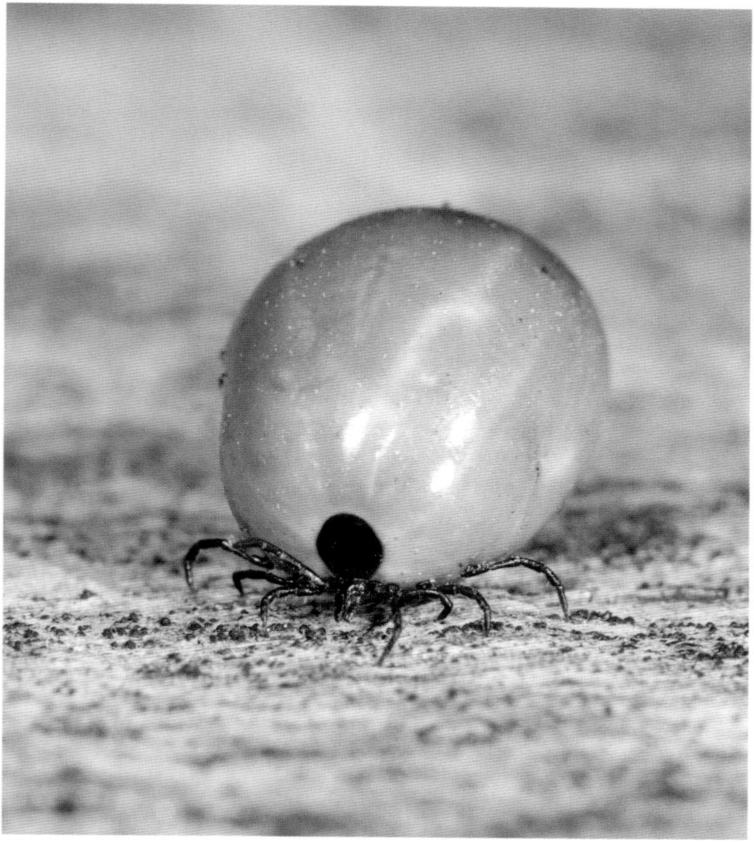

ALL PHOTOGRAPHS:

Castor bean tick

Also known as the sheep or deer tick, the castor bean tick inhabits humid woodlands and grasslands, feeding on the blood of large mammals, including humans. An engorged tick can be 11mm (0.43in) in size. Ticks spread a range of diseases, including Lyme's disease and encephalitis, which attacks the central nervous system.

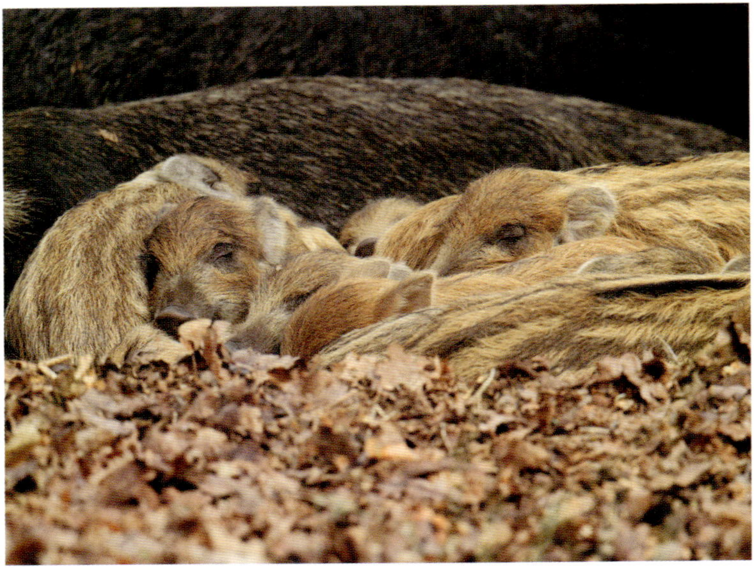

ALL PHOTOGRAPHS:

Wild boar

A common, dark-haired, inhabitant of oak and beech forests, the wild boar lives in groups of related females and their young. Males are solitary and have an average body length of 150cm (5ft). Its varied diet comprises roots, tubers, nuts, seeds, eggs, fish, molluscs and shoots. It also eats carrion, earthworms and small rodents.

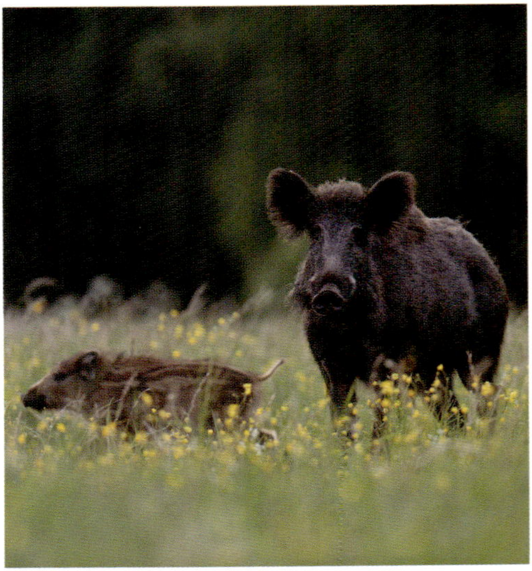

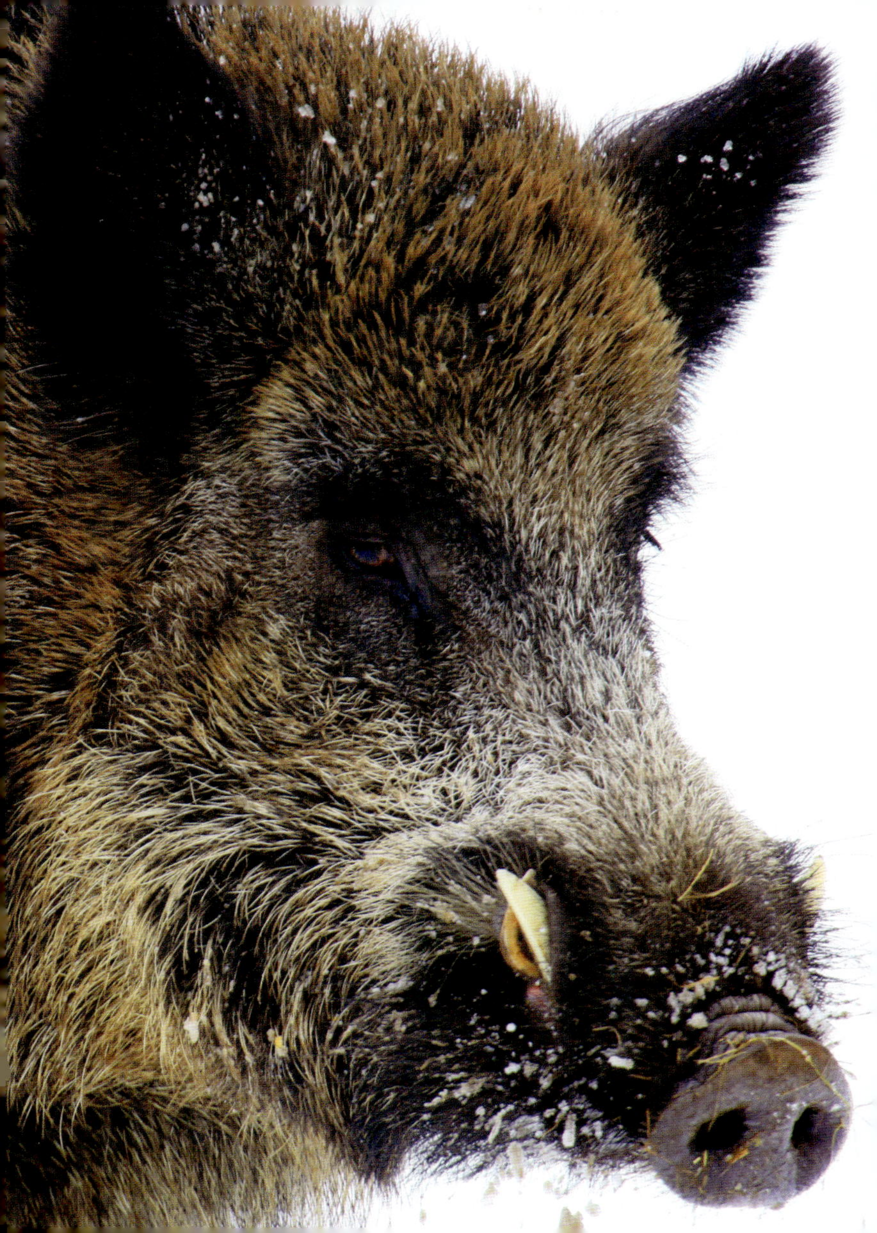

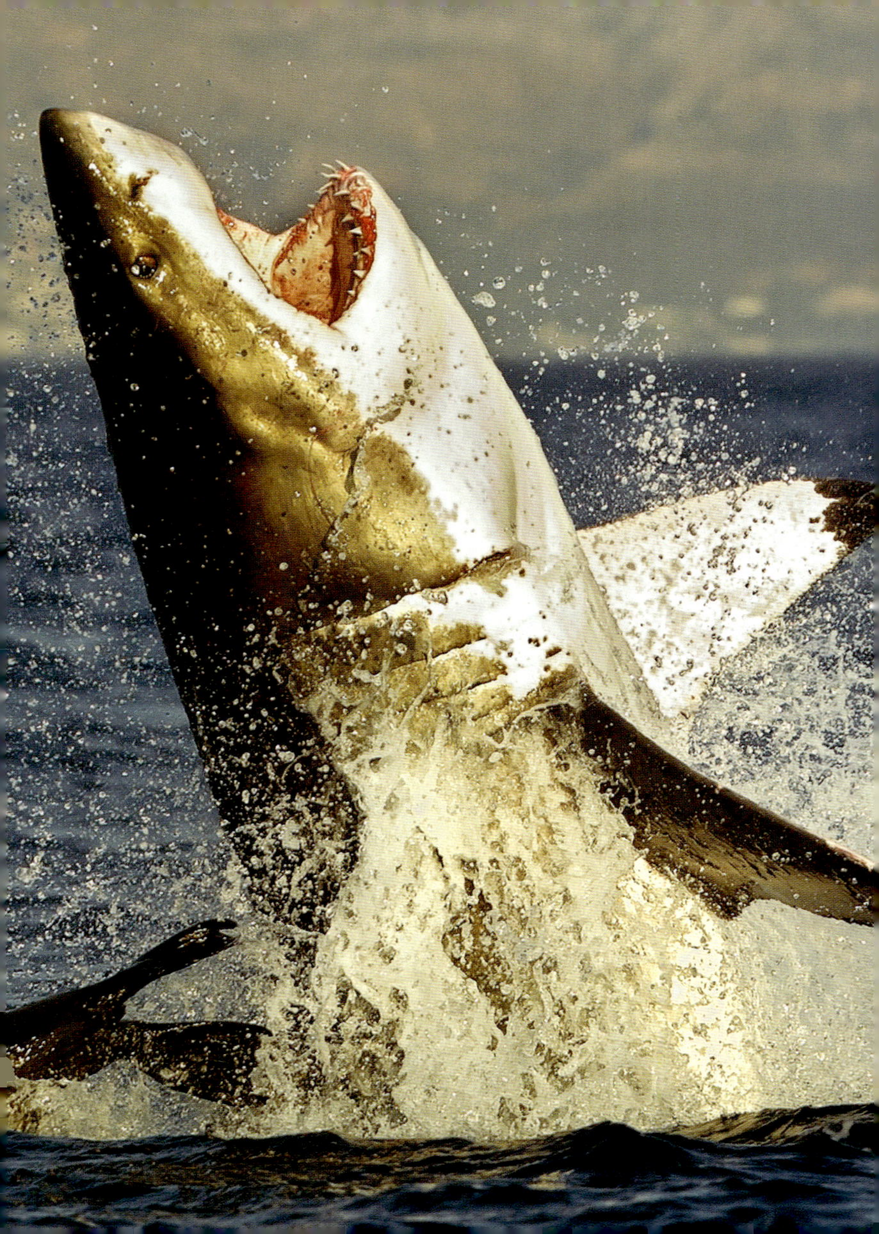

The World's Oceans

The oceans of the world make up over 70 per cent of the planet's surface. They are home to a vast range of animals; some are familiar and benign, others more ethereal. Its depths are home to many creatures that have featured in literature and myth and have become the subject of nightmares, as well as other unfamiliar organisms, so weird and wonderful that they almost seem part of some fantasy. There are some 230,000 cataloged species of animals and plants in our oceans. Among this throng can be counted the largest living organism, the immense blue whale; fish that can 'fly'; octopus and squid that change colour and form at need; and billions of microscopic organisms that glow with bioluminescence that is visible from space. Scientists believe there are many more species, maybe millions, awaiting discovery in the cold reaches of the deep. This chapter showcases some of the ocean's apex predators, those whose size, speed, agility, intelligence and power are such that they are unchallenged in their domains. Some, like the great white shark, will be old acquaintances; others, such as the lionfish, may be less well known but are equally as deadly. All play out their lives in an environment alien to us, but form part of an intricate web of life, bound together as part of a fragile ecosystem more prolific and bounteous that anywhere else on Earth.

OPPOSITE:

Great white shark

The largest predatory fish, with females commonly growing to 6m (20ft), the great white has obtained legendary status thanks to its portrayal in literature and film.

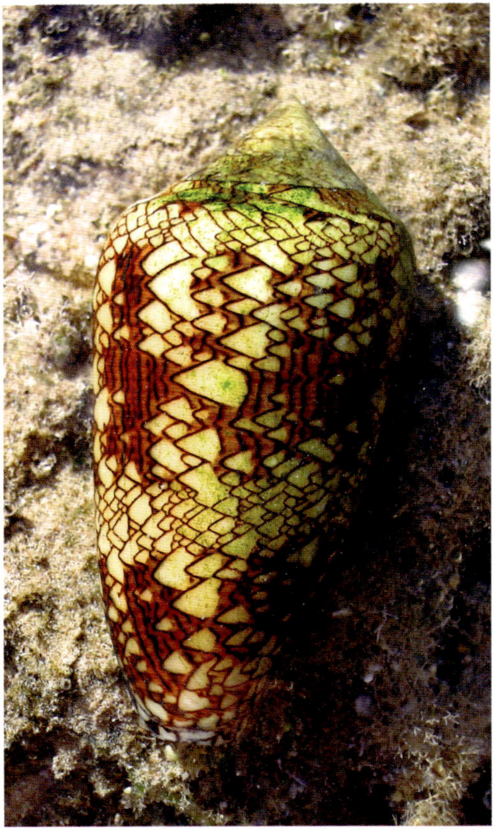

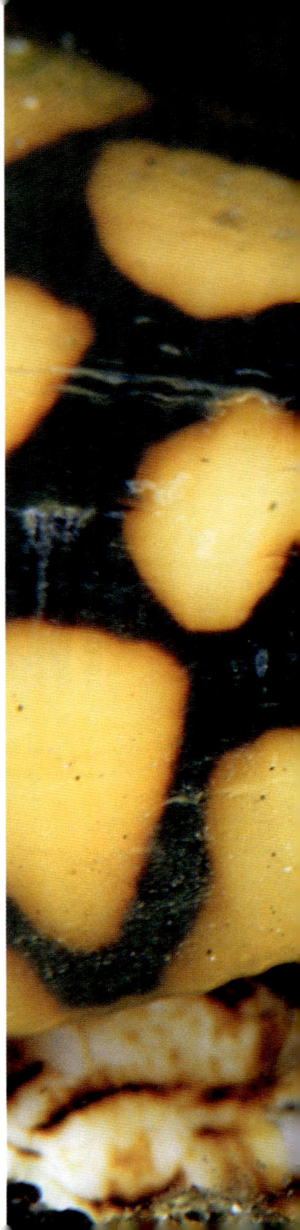

ALL PHOTOGRAPHS:

Cone snail

There are several hundred species of the predatory cone snail, the majority of which inhabit tropical waters. Many species have attractive, colourfully patterned shells, but should be handled with care. A barbed, needle-like 'tooth' is capable of injecting toxic venom that can cause painful injury and, in rare cases, death.

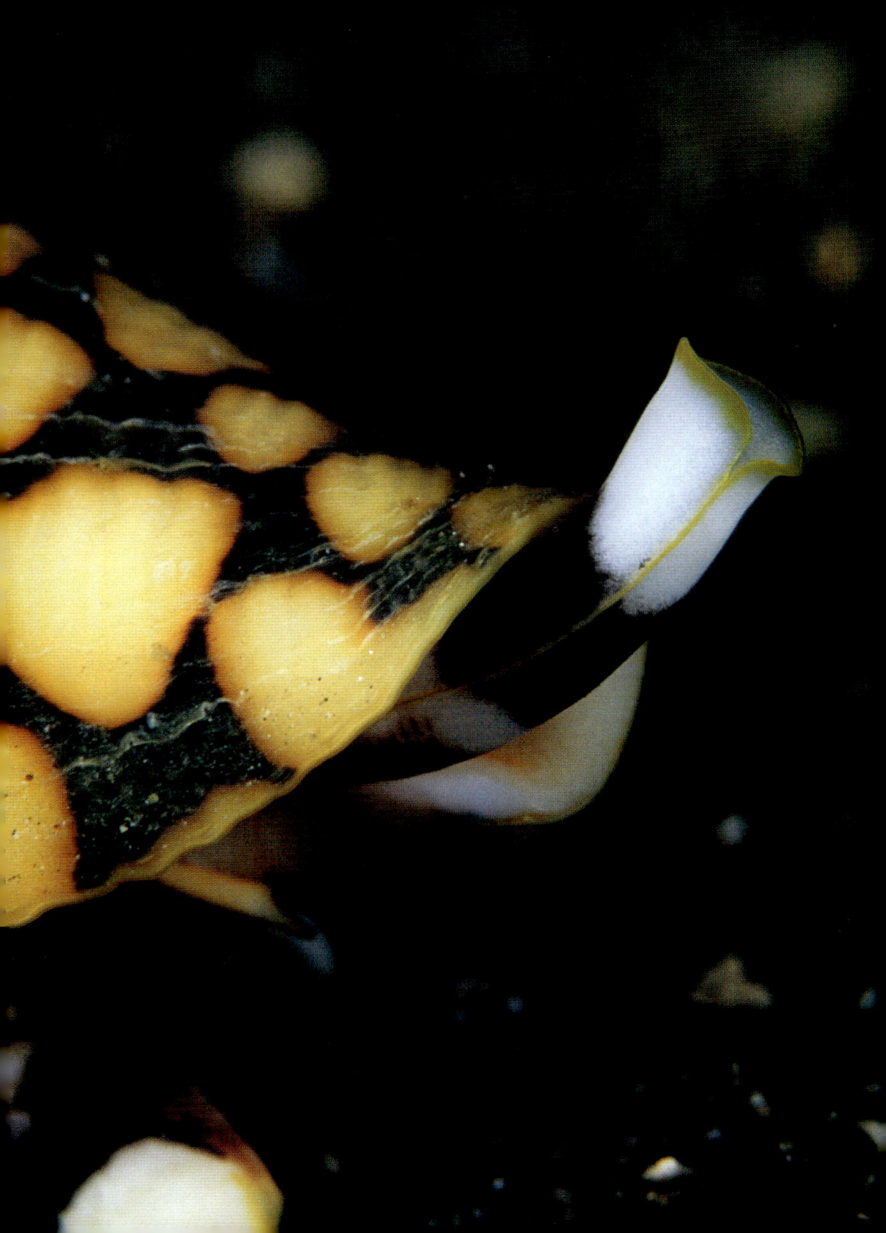

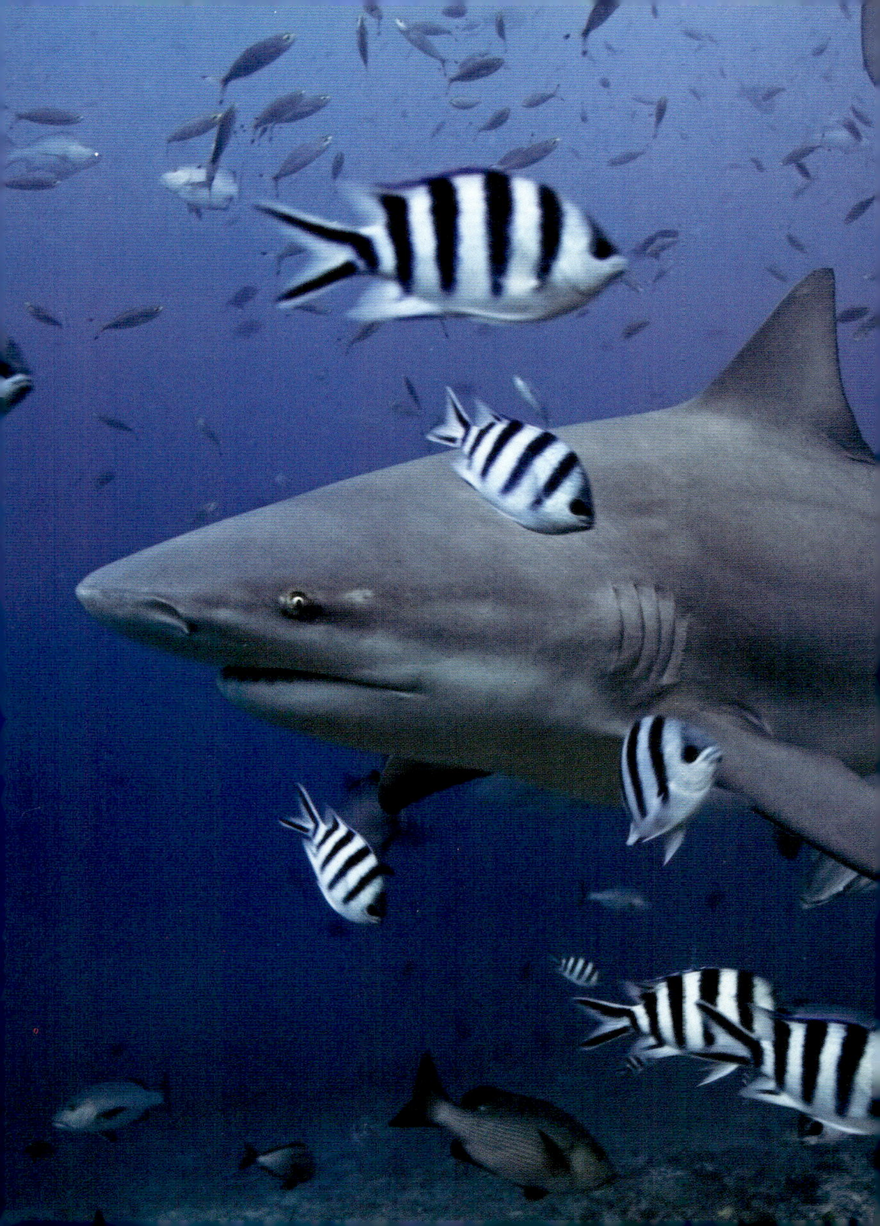

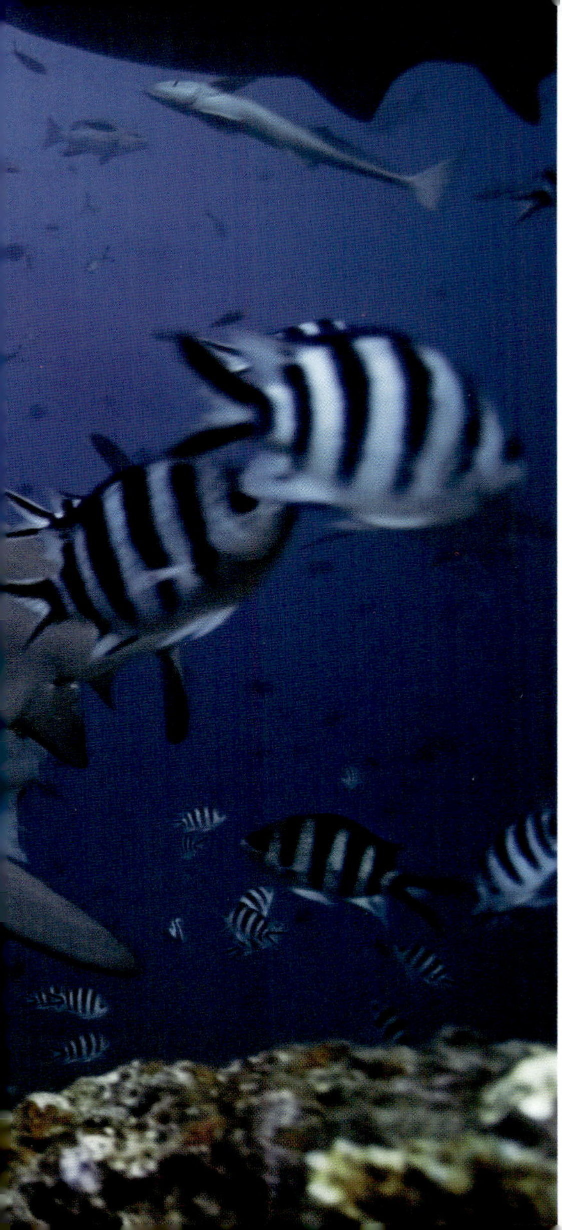

Bull shark

Bull sharks are found throughout warm oceans of the world. They are an unpredictable and aggressive species, recognised by their compact shape and broad snout. Their topside is grey and the underside white. They are unusual amongst sharks in that they are able to tolerate freshwater habitats.

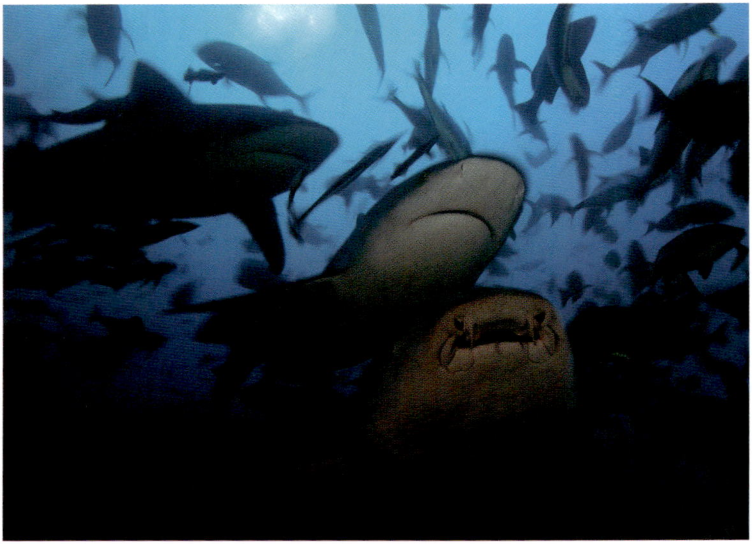

ABOVE AND RIGHT:

Bull shark

Bull sharks are large, with females averaging 2.4m (8ft) long; males are slightly smaller. They feed on smaller sharks, fish, birds, crustaceans and turtles, but are also one of the species most likely to bite humans. That said, humans are the biggest threat to bull sharks through overfishing for their meat.

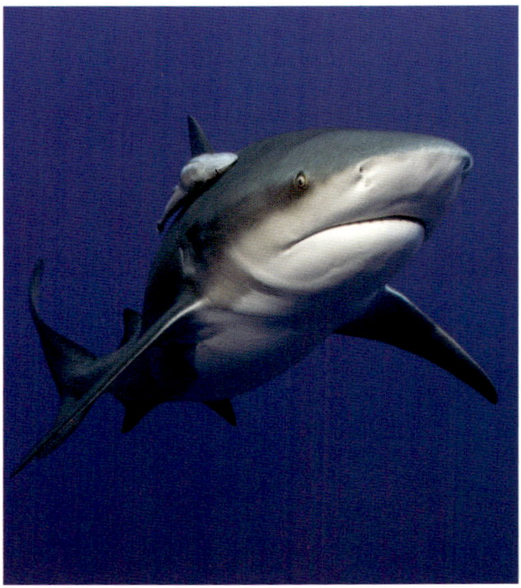

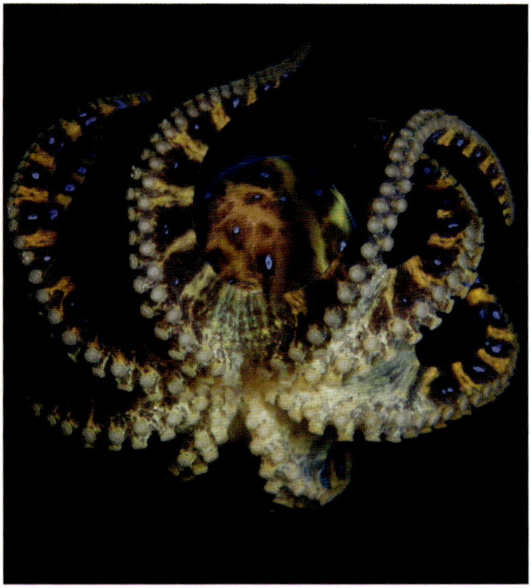

LEFT AND BELOW:

Blue-ringed octopus

Four species of this highly venomous octopus inhabit the coral reefs of the Pacific and Indian Oceans. They range in size from 12–20cm (5–8in), and are brightly coloured with a series of black-and-blue rings adorning their bodies and tentacles. When threatened, these rings pulse bright blue as a warning display.

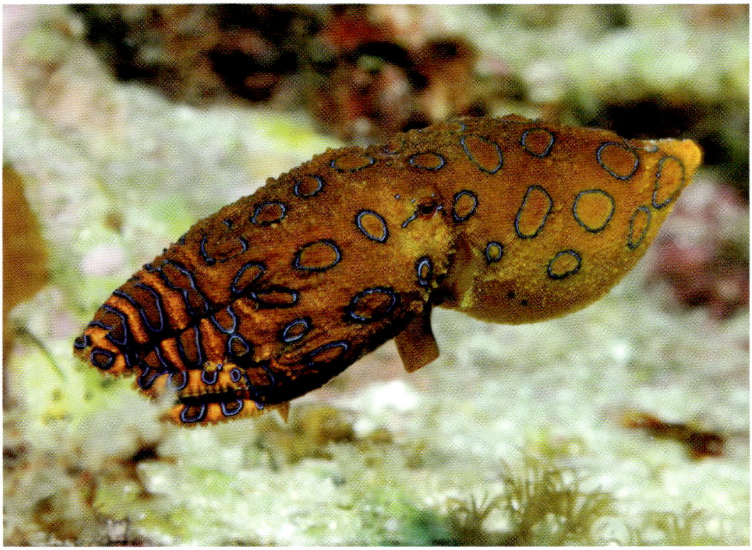

Blue-ringed octopus
This is the octopus of James Bond fame, administering the coup de grâce to his adversary in Ian Fleming's original Octopussy story. If handled roughly they are capable of injecting an extremely powerful neurotoxin which, if left untreated, is capable of killing a human within a few minutes.

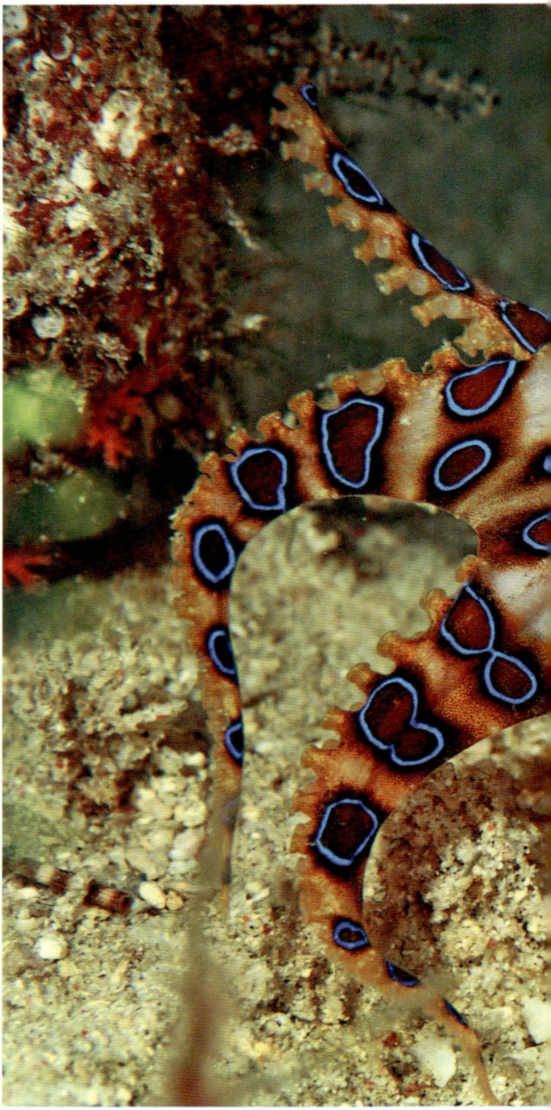

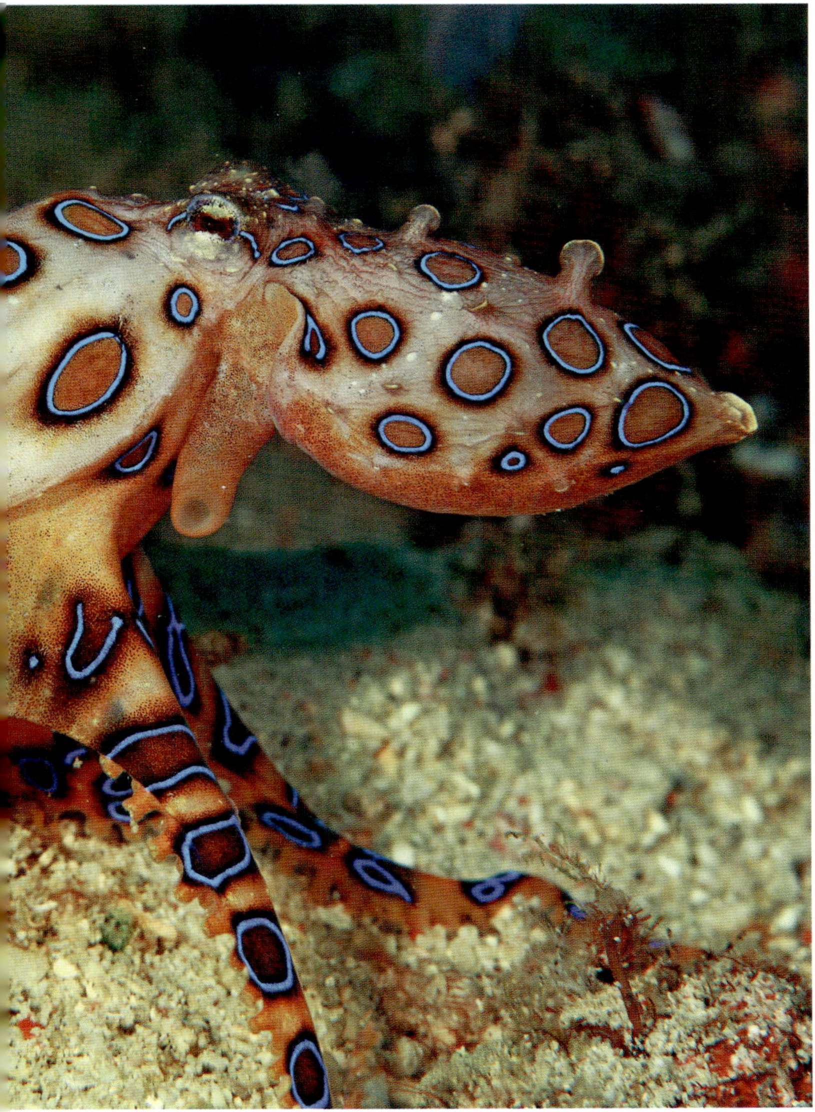

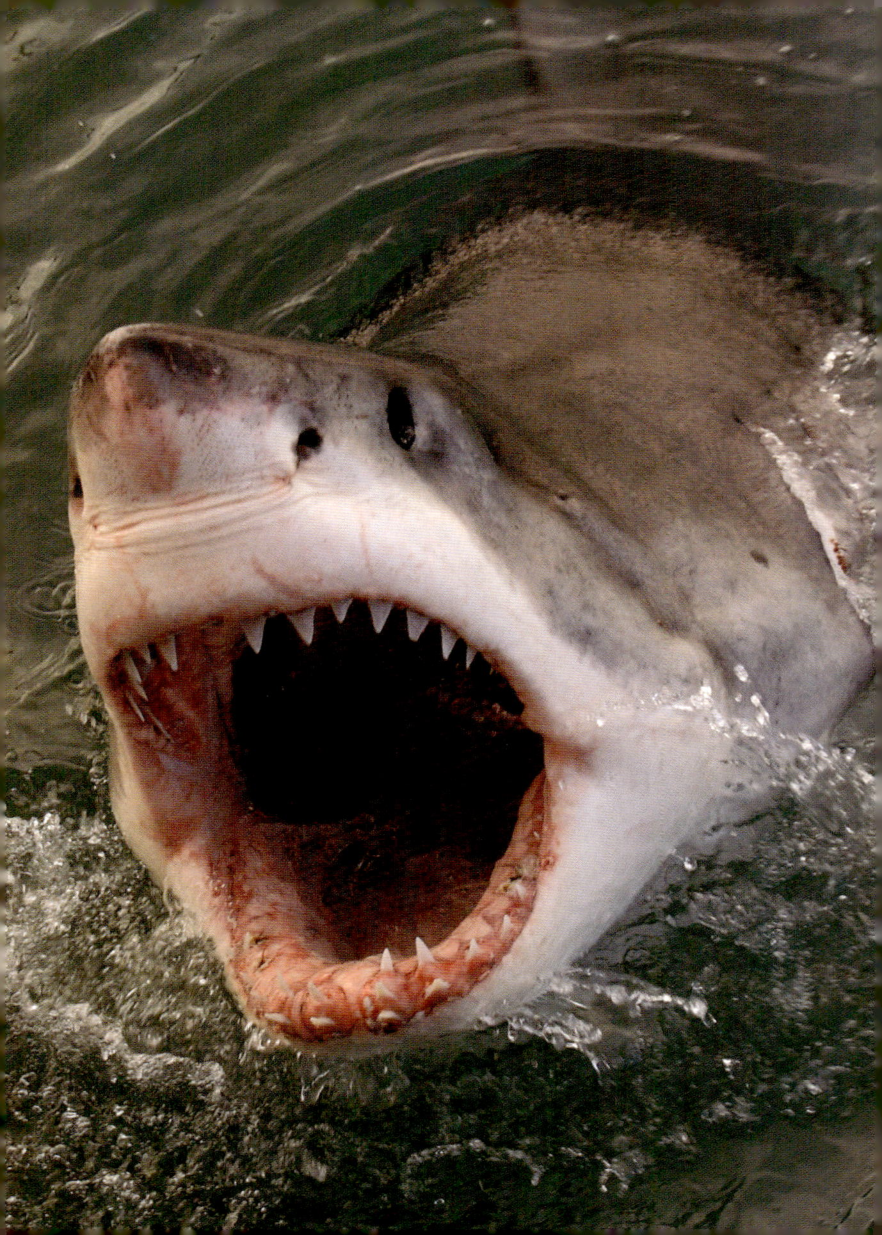

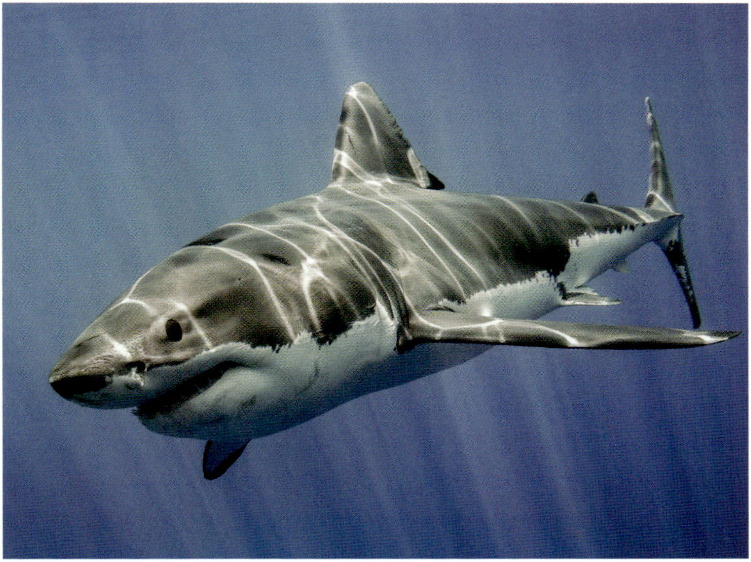

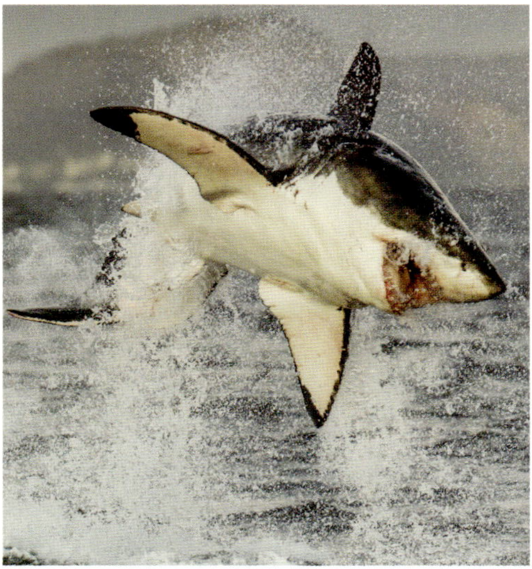

ALL PHOTOGRAPHS:

Great white shark

The great white has huge jaws full of razor-sharp, triangular teeth used to rip flesh. These sharks feed on fish, dolphins, turtles and sea mammals, ambushing seals from below and hitting them at high speed. During these attacks, sharks can reach speeds of up to 40km/h (25mph), their momentum carrying them clear of the surface.

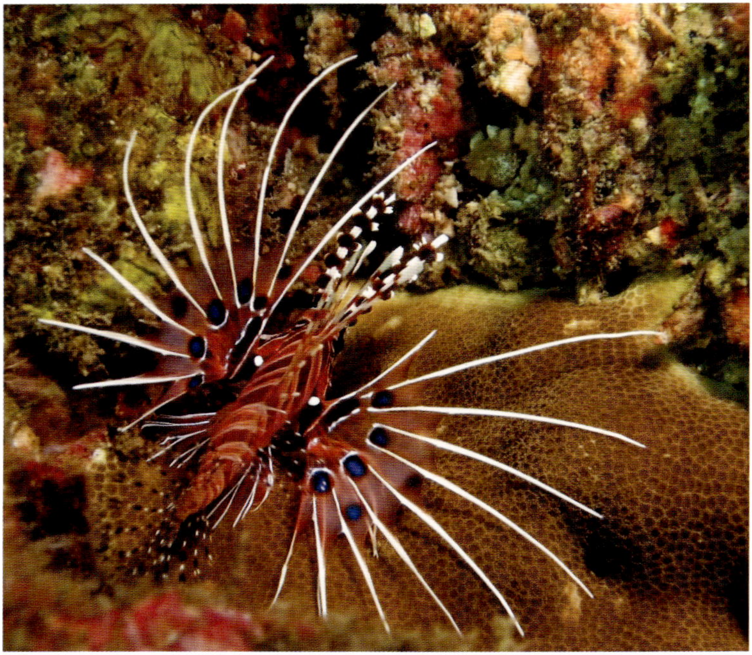

ALL PHOTOGRAPHS:
Lionfish
The lionfish is a group of distinctively shaped and patterned venomous fish, native to the Indian and Pacific Oceans. It can grow to 50cm (1.5ft), and administer venom via its long spines. It has become an undesirable invasive species outside its native range, where its voracious appetite represents a major threat to fragile ecosystems.

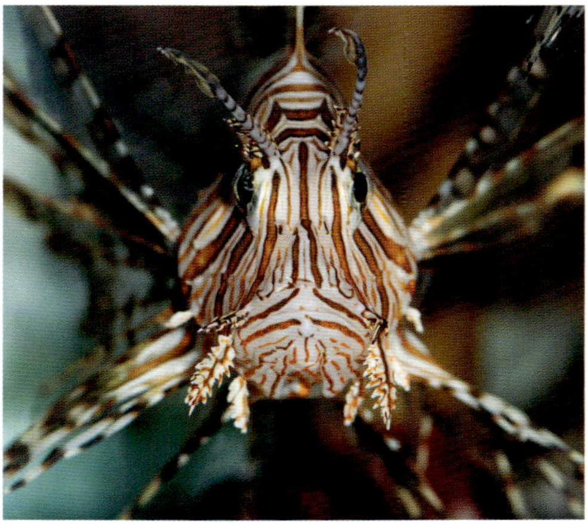

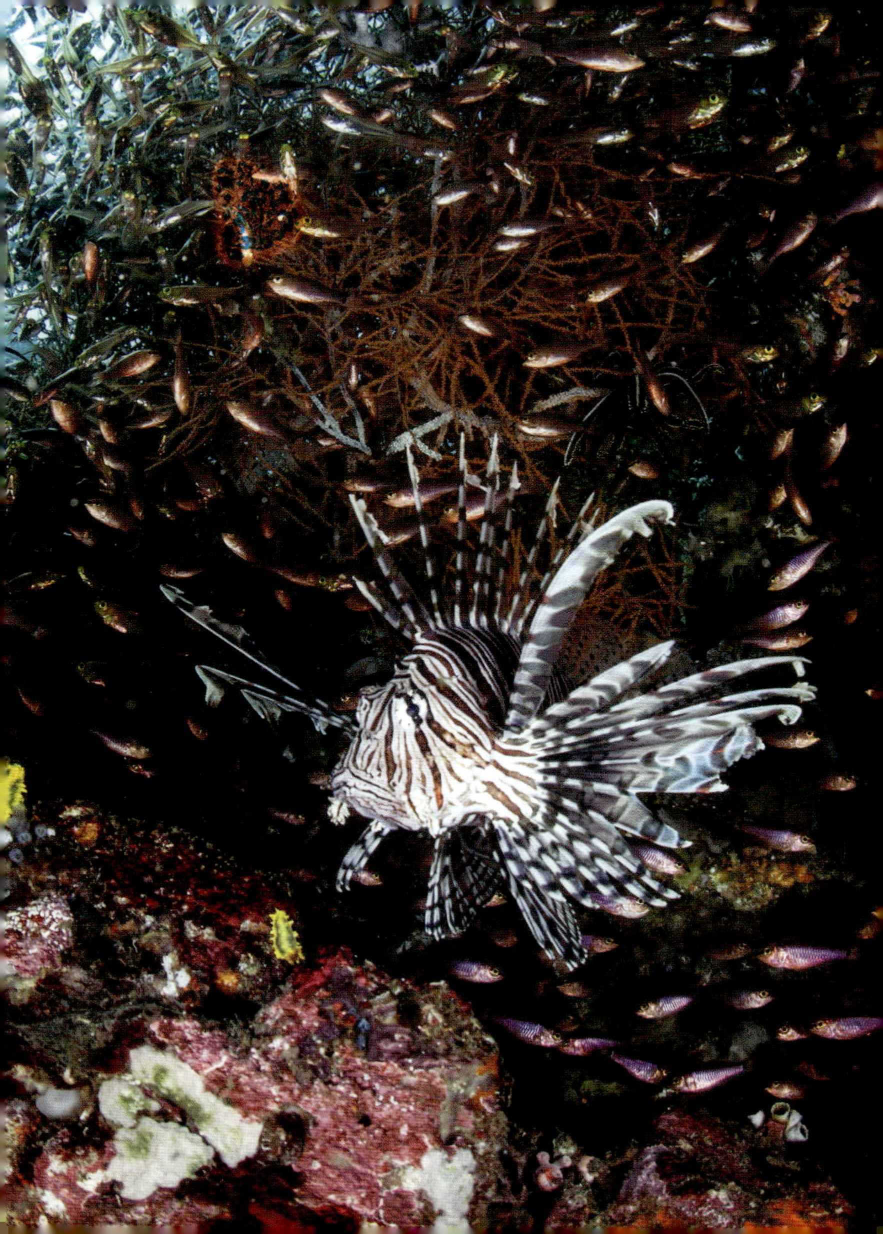

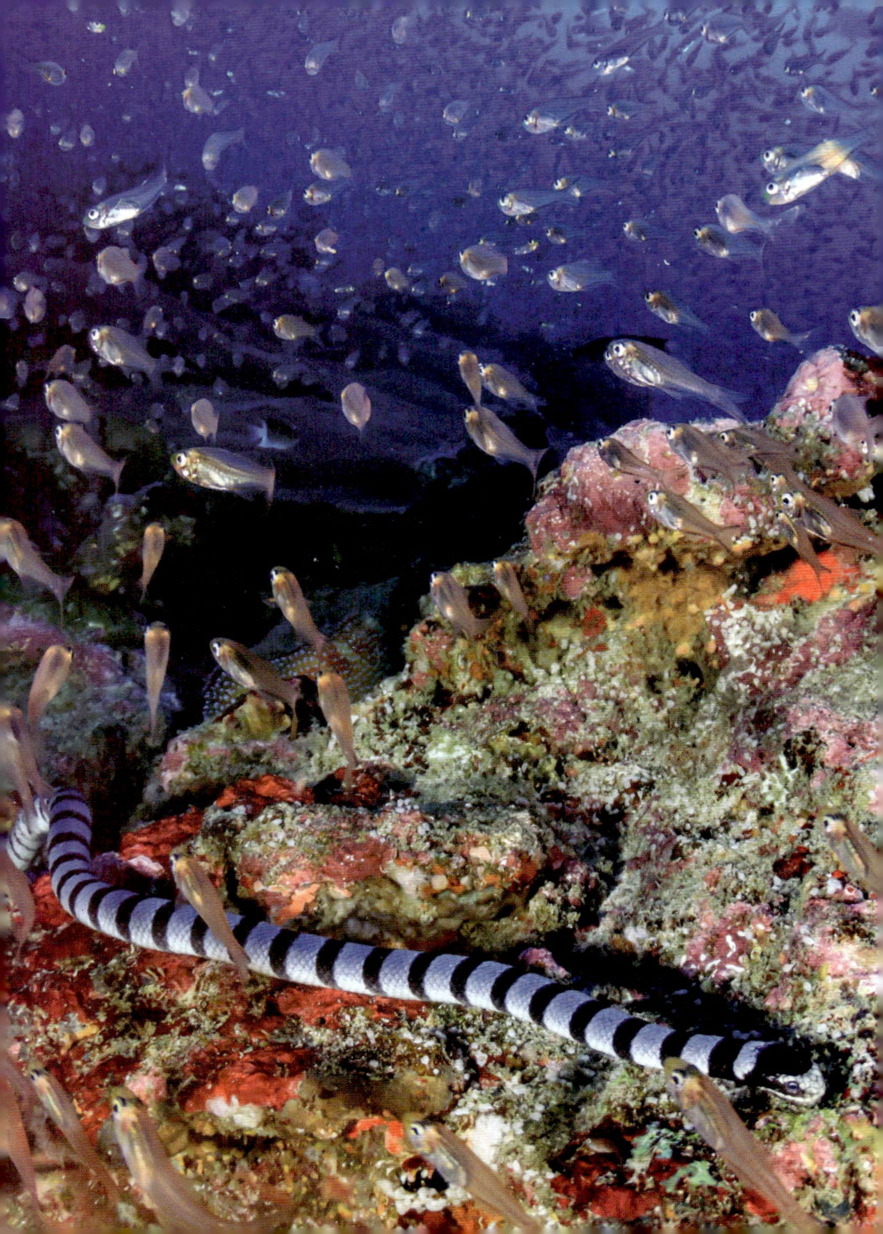

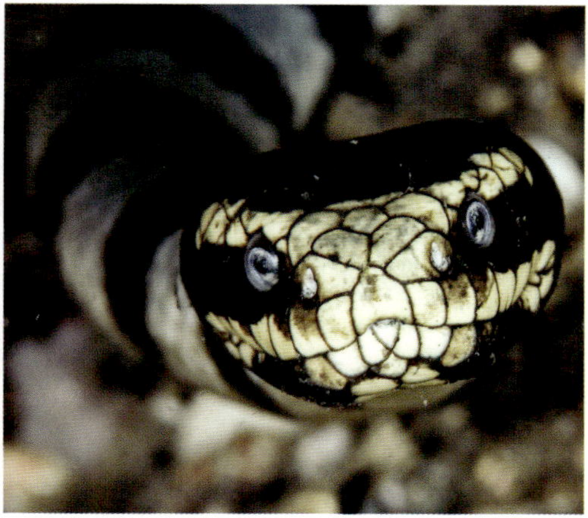

ALL PHOTOGRAPHS:

Sea snake

A group of colourful, fully aquatic snakes with paddle-like tails to aid swimming, sea snakes can grow to a length of 150cm (5ft) and inhabit warm waters of the Indian and Pacific oceans, where they hunt small fish. Although some species are highly venomous, they are not generally aggressive.

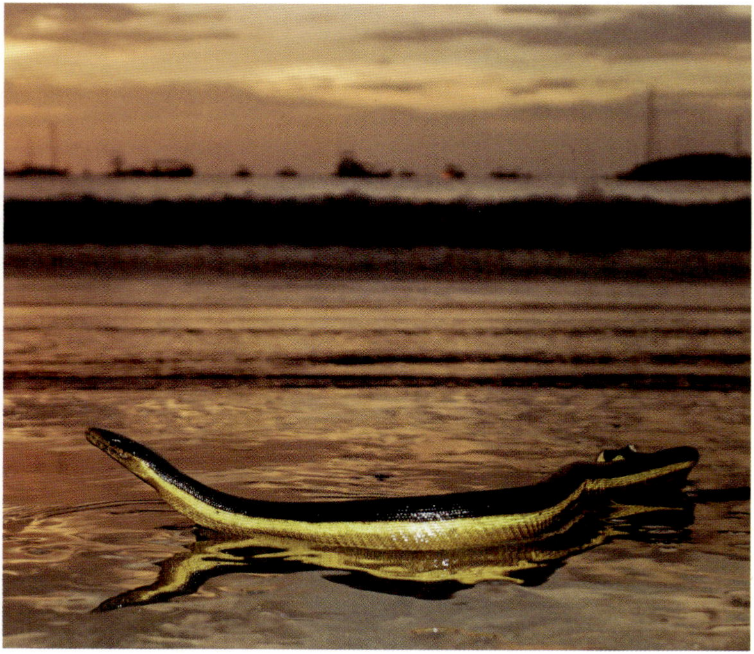

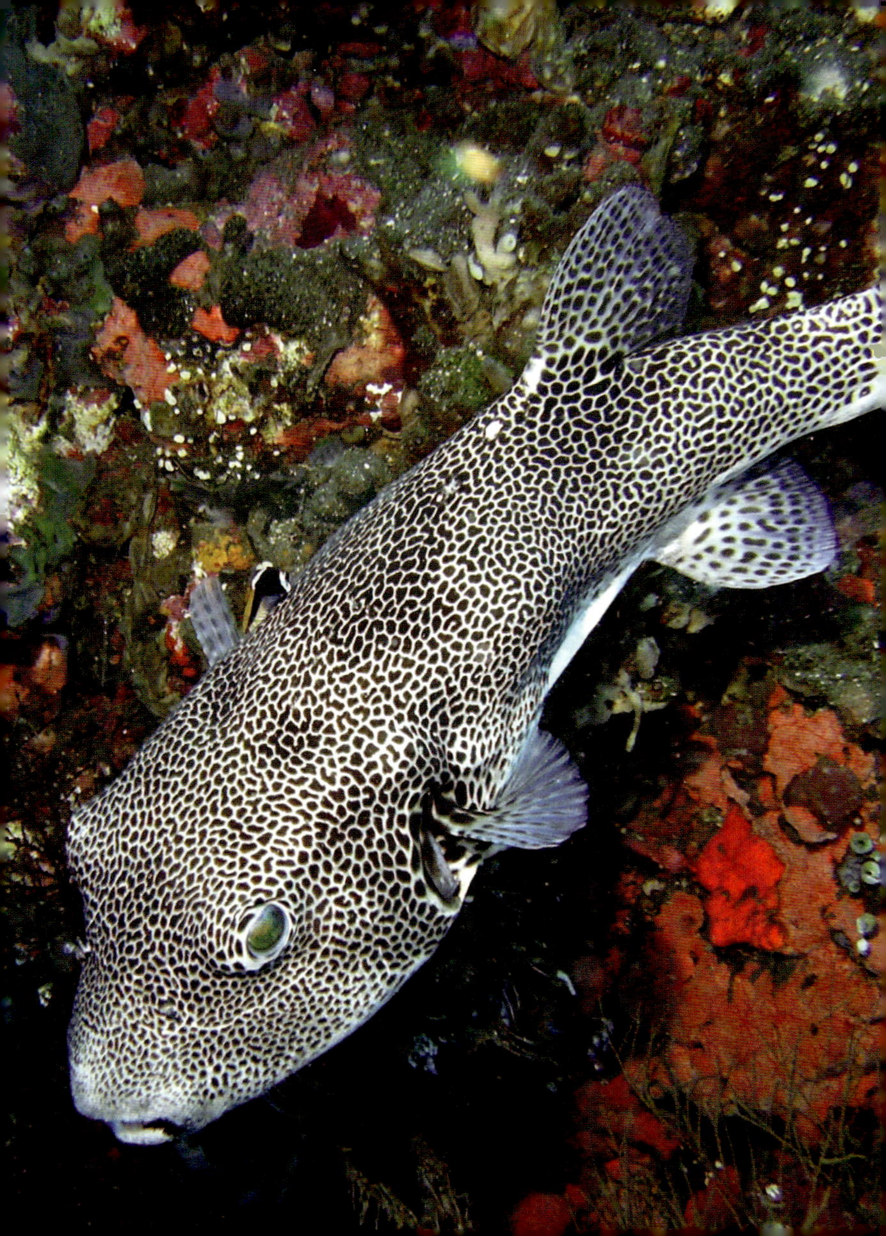

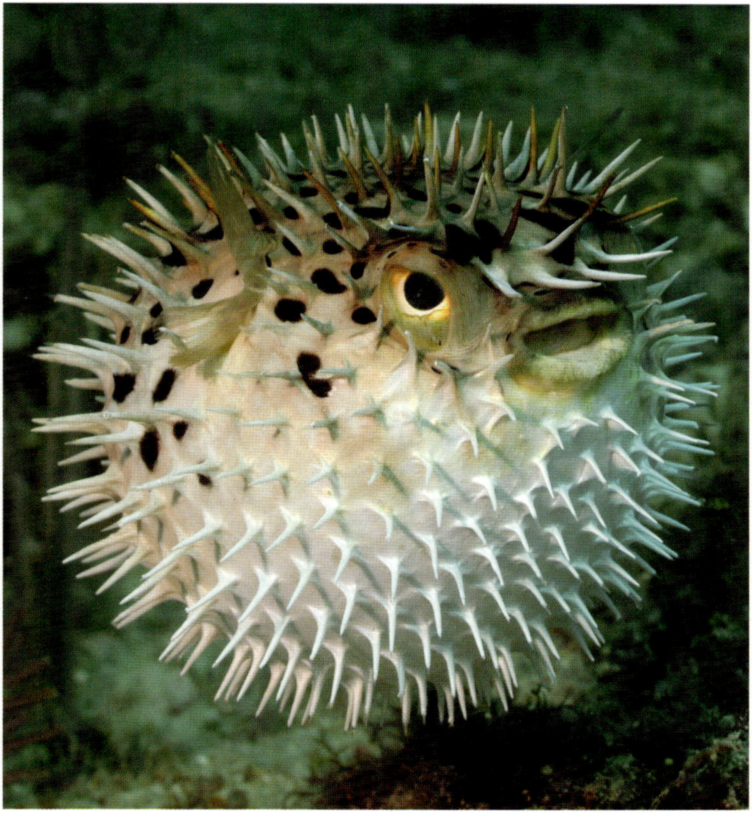

ALL PHOTOGRAPHS:

Puffer fish

A defence mechanism of this 50cm (1.6ft) long fish is to inflate its stomach with air or water, making it almost spherical and therefore less palatable to predators. All species also have sharp spines that help to deter any aggressor. The skin and internal organs of most species are highly toxic.

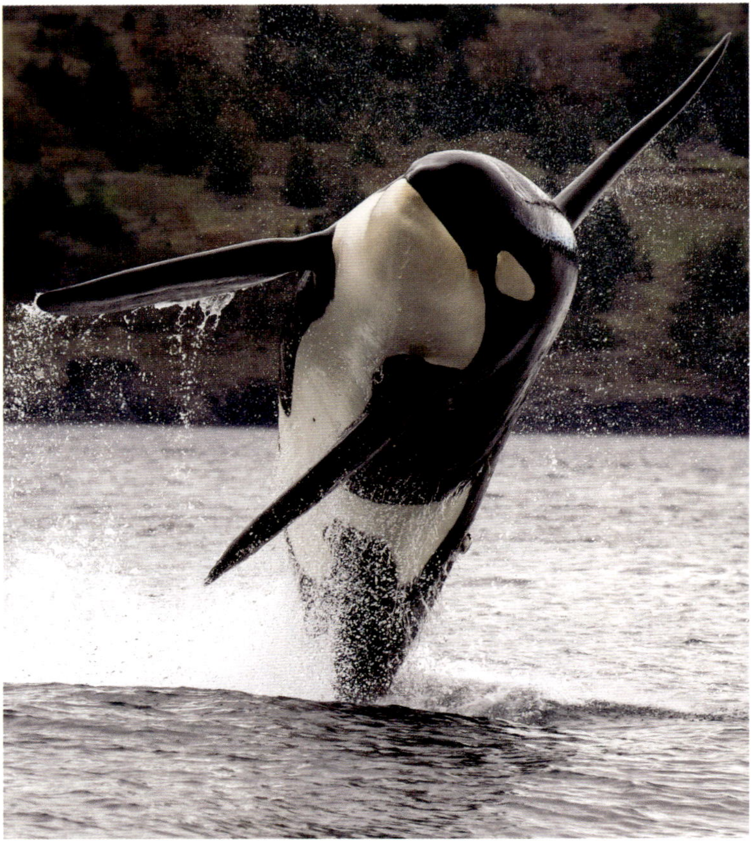

ABOVE:

Orca male breaching

Orcas, or killer whales, are the largest members of the dolphin family. Mature males grow to 8m (26ft) and have long, pointed dorsal fins up to 1.8m (6ft) long. Females are smaller and have a much reduced, curved dorsal fin. They are found in all the oceans of the world.

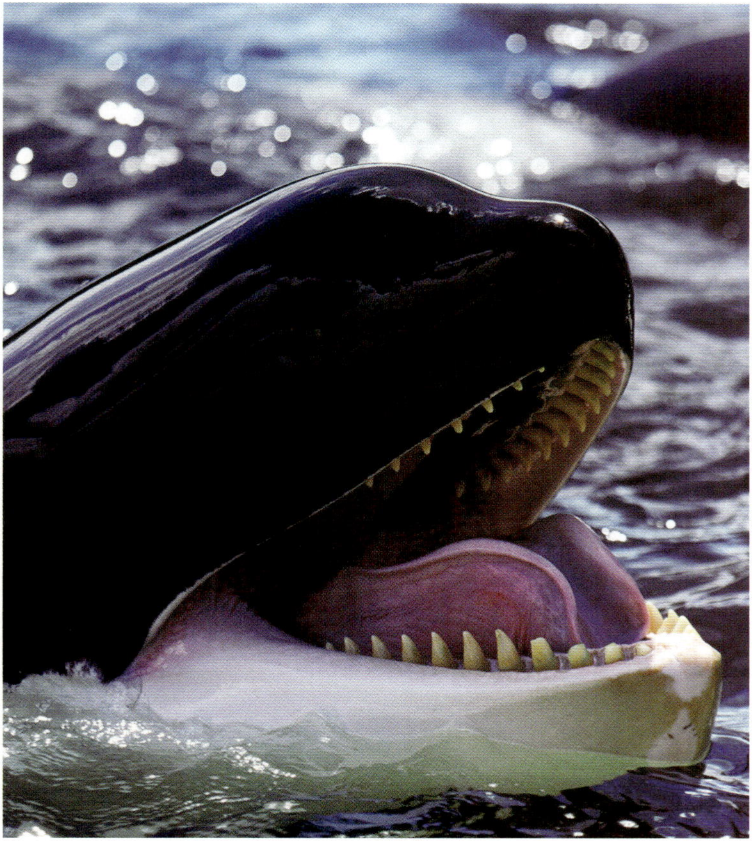

ABOVE:

Orca

Orcas have strong, sharp teeth used to grasp and dismember their prey, which comprises fish and marine mammals including dolphins and whales, and birds. They hunt in packs, employing their excellent eyesight and hearing to locate food. They are intelligent animals, able to learn complex hunting techniques from older animals.

Pod of orcas

Orcas are long-lived animals with a complex social structure. They live in family groups called pods containing members of several generations. They are highly vocal, communicating with members of their own pod and others by sophisticated vocalisation. Although labelled 'killers', there are no recorded fatal attacks on humans from wild orcas.

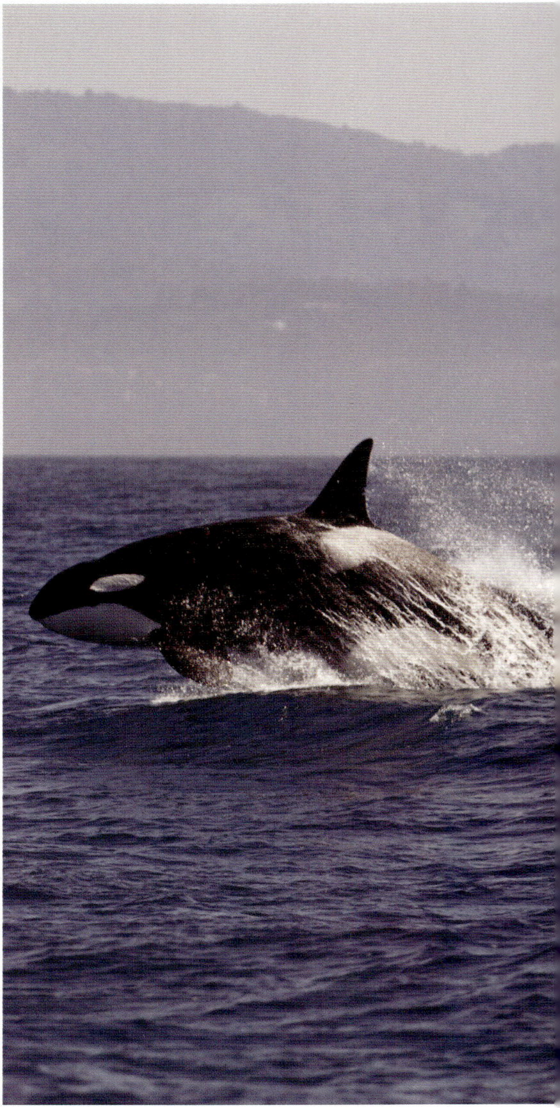

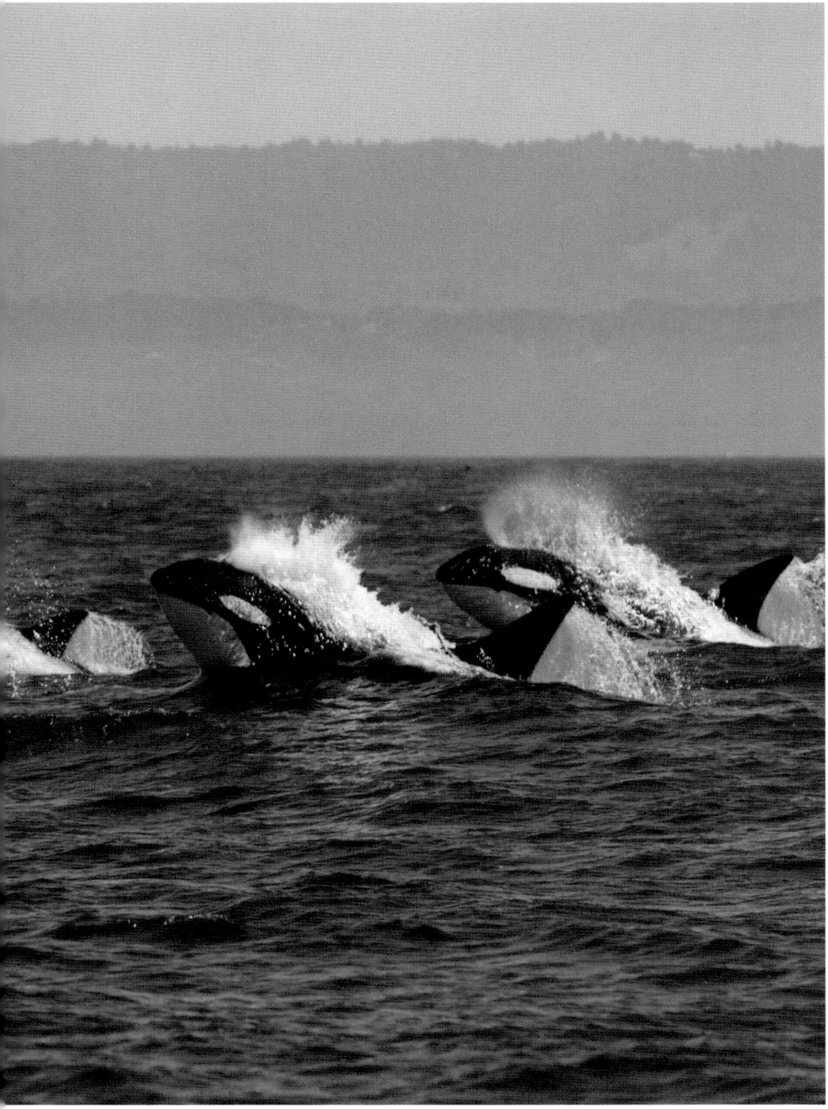

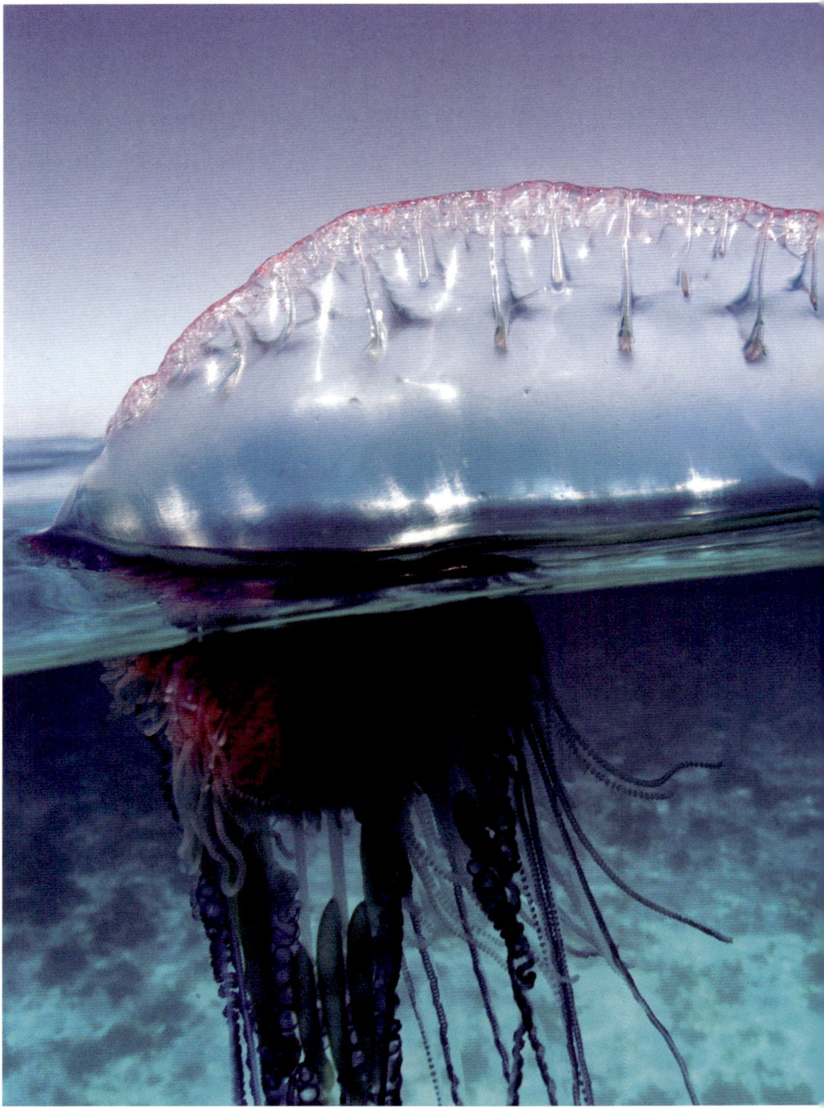

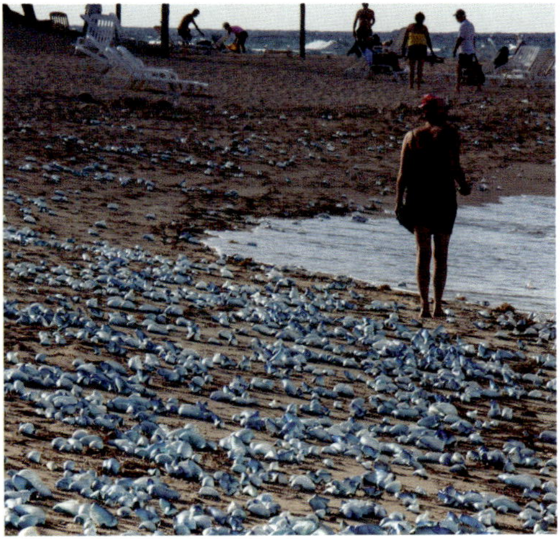

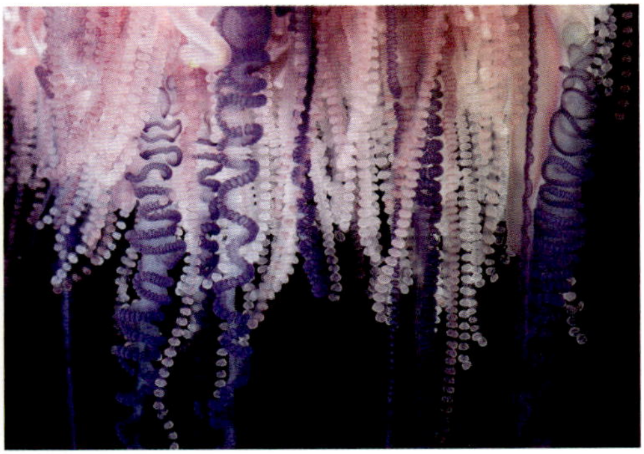

ALL PHOTOGRAPHS:

Portuguese man o' war

This jellyfish is a colonial organism made up of many genetically identical components, each having a discrete function. The sail like 'bladder' is the most obvious component, but the trailing 10m (33ft) tentacles are highly venomous, and used to stun prey such as fish and shrimps. This sting is very painful to humans, but rarely fatal.

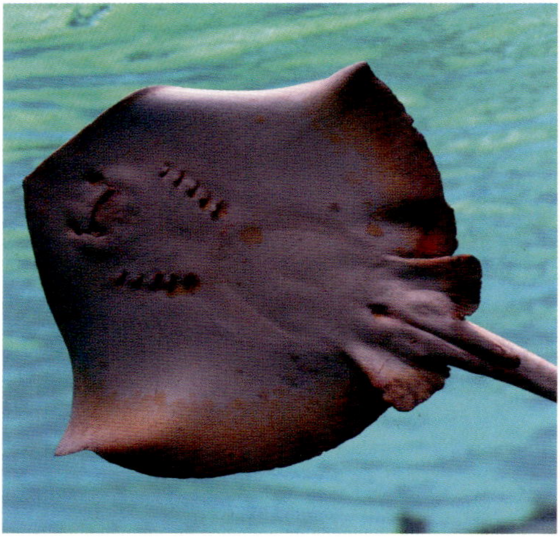

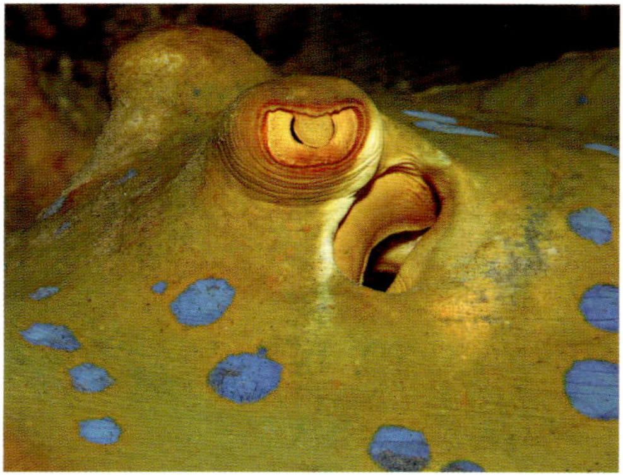

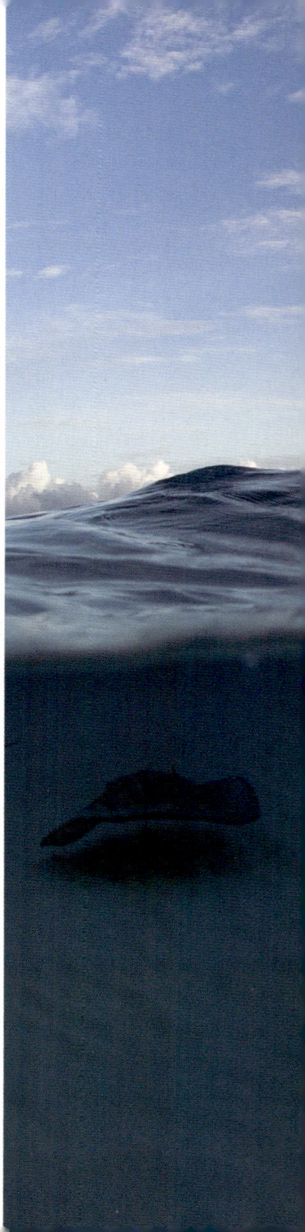

ALL PHOTOGRAPHS:

Stingray

There are over 200 species within this group inhabiting warm, tropical oceans where they feed on plankton, crustaceans, molluscs and fish. They hunt by foraging or ambush, some using their strong teeth to crush prey. The 'sting' is used for defence, and although it causes severe pain, it is rarely fatal.

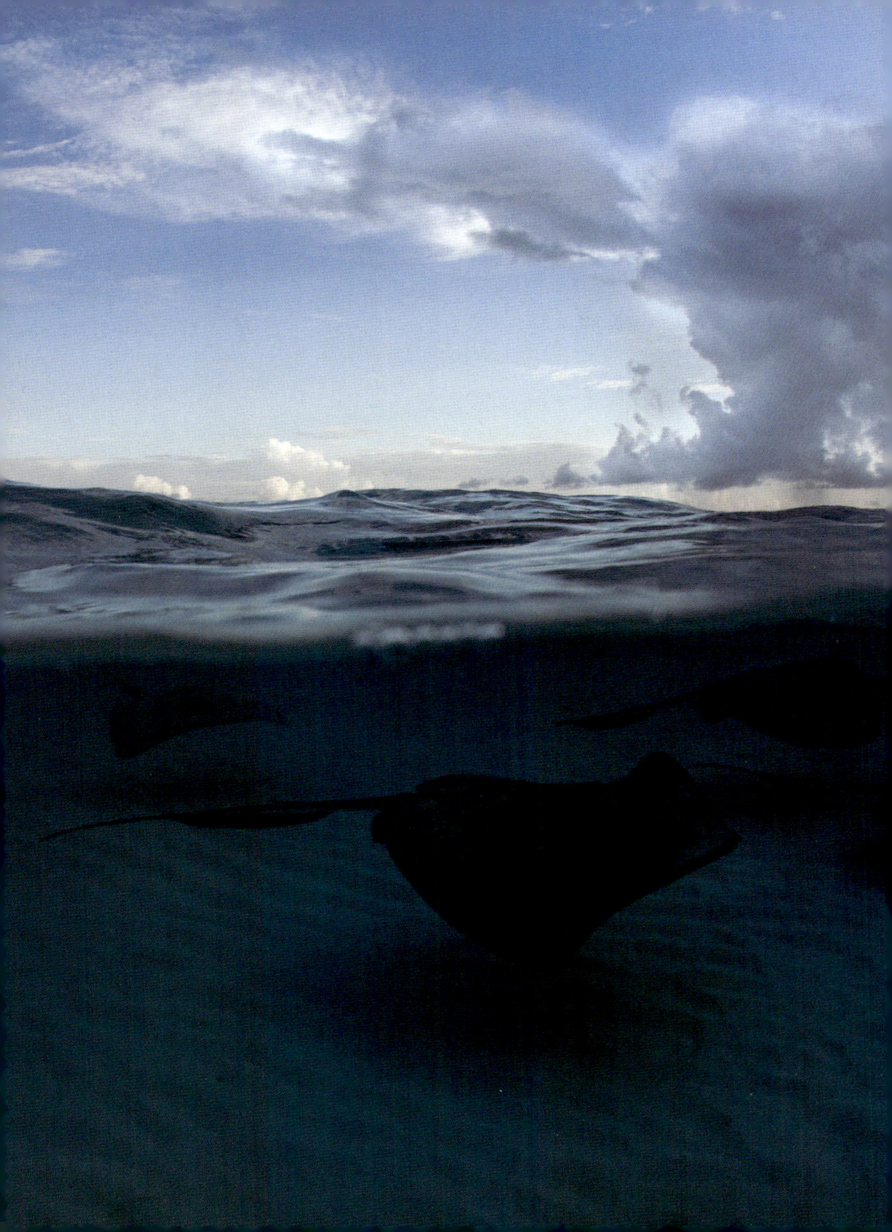

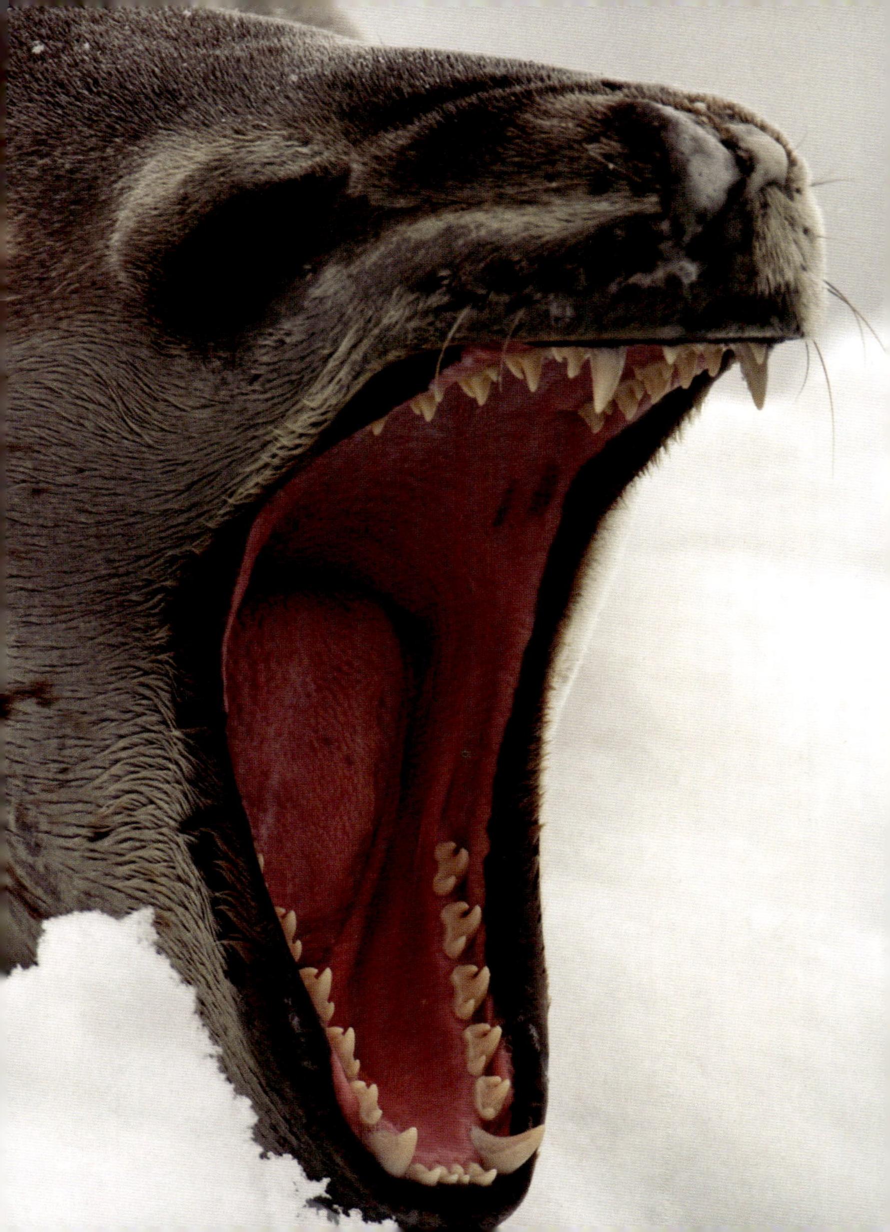

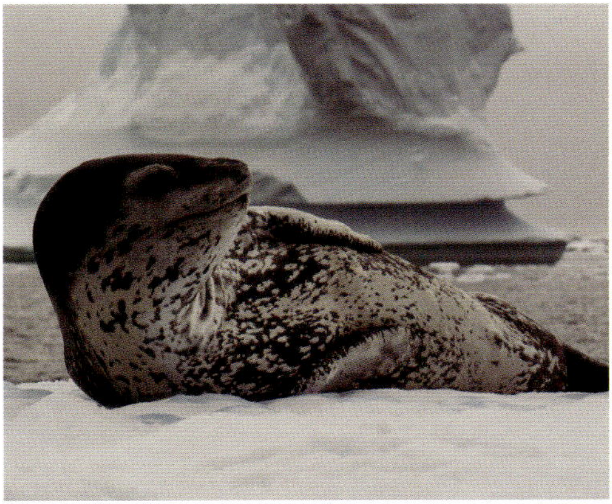

ALL PHOTOGRAPHS:
Leopard seal
Their powerful jaws, with canine teeth 2.5cm (1in) long, allow leopard seals to prey on smaller seals and penguins, which they hunt at the edge of ice flows. They pose little threat to humans, with only a single recorded instance of a fatal attack, probably due to the diver being mistaken for a fur seal.

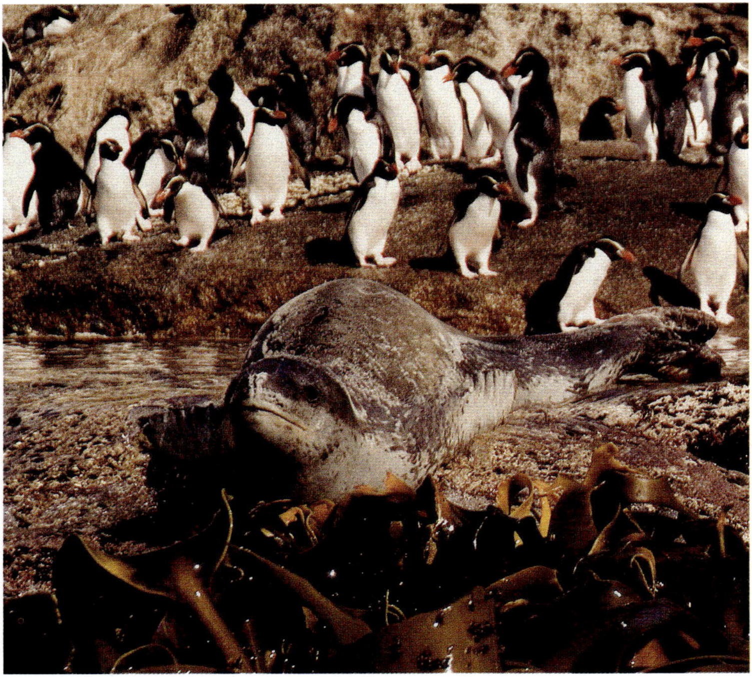

Picture Credits

Alamy: 10/11 (Martin Harvey), 14/15 (Blaize Pascall), 16 top (Matthijs Kuijpers), 19 bottom (imageBroker), 30/31 (AfriPics.com), 34/35 (Brian Pollard), 49 bottom (Panoramic Images), 50/51 (Helmut Corneli), 63 (Dinodia Photos), 72 top (Auscape International Pty Ltd), 78/79 (Nature Picture Library), 80 (Jonathan Ayres), 82/83 (Peter Yeeles), 88/89 (Jonathan Ayres), 90 top (Ken Hawkins), 90 bottom (David Kleyn), 91 (Paul Rollins), 92 top (Barb Giles), 93 (AlphaCharlie Photography), 110 bottom (John Cancalosi), 111 top (Matthijs Kuijpers), 116 (Eyal Bartov), 117 top (Geoff du Feu), 117 bottom (Imagebroker), 118/119 (Judy Yaker), 121 (Ray Bulson), 138 (James Harrison), 139 bottom (Redmond O. Durrell), 141 top (Minden Pictures), 141 bottom (Danita Delimont), 142/143 (Minden Pictures), 143 top (robertharding), 143 bottom (Imagebroker), 146 (rod Williams), 147 top (Les Gibbon), 147 bottom (blickwinkel), 148 (Nature Picture Library), 149 (Minden Pictures), 151 top (Octavio Campos Salles), 157 (Juniors Bildarchiv GmbH), 158 (Minden Pictures), 159 (imageBroker), 162/163 (George Grall), 164 top (adrian Hepworth), 164 bottom (Klaus Ulrich Muller), 165 (Matthijs Kuijpers), 166 (Nature Picture Library), 168 top (Pally), 171 (blickwinkel), 177 (mauritius images GmbH), 178 top (Nature Picture Library), 178 bottom (Arterra Picture Library), 182/183 (Minden Pictures), 186/187 (Alamy Stock Photo), 193 (Hans Verburg), 194 top (Lewis Thomson), 199 (WaterFrame), 203 top (James Peake), 204/205 (imageBroker), 209 (Stocktrek Images Inc), 218 (Stephen Frink Collection), 219 bottom (Nature Picture Library), 223 bottom (Imagebroker)

Dreamstime: 7 (Ramon Carretero), 16 bottom (Poco bw), 17 (Dirkwenzel), 18 top (Ecophoto), 20 top (Sloth92), 20 bottom (Silksatsunrise), 21 (Lucasdm), 22/23 (Stuporter), 24 (ePhotocorp), 25 (Surz01), 28 bottom (Artushfoto), 29 (Hphimagelibrary), 32 (Ondrejprosicky), 36 (Leighdiprosephotography), 37 (Kbraunsd), 38/39 (Ecophoto), 40 top (Johan63), 40 bottom (Ecophoto), 41 (Tjkphotography), 42 (Byrdyak), 44 top (Foto76), 45 (Saranvaid), 46/47 (Sourabhis), 48 (Jenslambertphotography), 49 top (Utopianepal), 52 (ePhotocorp), 53 (Mgkuijpers), 54 (Jxpfeer), 55 top (Wrangel), 56 top (Bobonacus), 56 bottom (Curioustiger), 56/57 (Photomaru), 58/59 (Surz01), 60 (Mgkuijpers), 61 (Neilneil), 62 top (Sulemani1967), 62 bottom (Voraphong), 64/65 (Rgbe), 66 top (Donvanstaden), 66 bottom (Stuporter), 67 (Altaoosthuizen), 68/69 (Delonybro), 70 (Kengriffiths6), 73 (Lessysebastia), 76 top (Leitnerr), 76 bottom (Fejkl1), 77 top (Catchyimages), 77 bottom (Lunatic67), 78 (Ozflash), 84/85 (PeterWaters), 86 (Lucidwaters), 87 top (Kengriffiths6), 87 bottom (Dwiputra18), 92 bottom (Dennisjacobsen), 94/95 (Kengriffiths6), 98/99 (Viktoriart), 100 (Pixelstate), 102/103 (Gkuchera), 104 (Gnagel), 105 (Rickmcmillin), 106/107 (Simonwehner), 108 top (Naluphoto), 108 bottom (Mike711), 109 top (Isselee), 109 bottom (Asterixvs), 110 top (Tazzy), 111 bottom (Kwiktor), 112/113 (Windowsontheworldphotography), 113 top (Neilld), 113 bottom (Silksatsunrise), 114 (Byrdyak), 115 (Andreanita), 120 bottom (Davidstockphoto), 122/123 (Mikaelmales), 124 (Denisapro), 125 top (Ondrejprosicky), 126 (Visceralimage), 127 (Erikmandre), 128 (Pxhidalgo), 130 (Viniciussouza06), 131 (Taciophilip), 132/133 (Jens Otte), 134 bottom (Blackmambafive), 135 (Ondrejprosicky), 136/137 (Arniearup), 139 top (Danolsen), 144/145 (Bookdreamerjulia), 150 (Dalia Kvedaraite), 151 bottom (Kwiktor), 154/155 (Lukyslukys), 156 (Buckskinman), 168/169 (Ondrejprosicky), 172 (Surz01), 173 (Rinofelino), 176 (Schmali666), 179 (Twildlife), 184 top (Empipe), 188/189 (Patrick), 190 (Gudkov2000), 191 bottom (Photosebek), 194 (Erikmandre), 195 (Jensklingebiel), 196 (Surz01), 198 (Djay712), 200/201 (Prochazkafoto), 202 bottom (Fionaayerst), 203 bottom (Oksanavg), 207 top (Ramoncarretero), 207 bottom (Surz01), 208 top (Pufferfishy), 210 (Whitcomberd), 211 bottom (Johnpierpont), 212 (Mherman), 213 (Allnaturalbeth), 220 bottom (Lilithlita), 222 (Kohavi), 223 top (Vladsilver)

Shutterstock: 8 (javarman), 12/13 (CherylRamalho), 19 top (Rudi Hulshof), 26 (Vladimir Turkenich), 27 top (Ari Wid), 27 bottom (Marcel Derweduwen), 28 top (Volodymyr Burdiak), 32/33 (mark chilton), 44 bottom (Anan Kaewkhammul), 55 bottom (FLPA), 72 bottom (Malcolm Schuyl), 74/75 (Jonathan Steinbeck), 81 (Uwe Bergwitz), 82 (D.M.Ritchie), 96/97 (Ken Griffiths), 120 top (Jolanda Aalbers), 125 bottom (Dennis Jacobsen), 133 (Uwe Bergwitz), 134 top (Jens Otte), 140 (Walter Mario Stein), 152/153 (Next is nicer-KR7), 160/161 (Andrey Armyagov), 168 bottom (Tony Baggett), 170 (Carlos Aranguiz), 174/175 (Erik Mandre), 180/181 (Vlad Sokolovsky), 184 bottom (GenOMart), 185 top (bekirevren), 191 top (Jan Stria), 192 (KPixMining), 202 top (Connect Images Curated), 206 (Alessandro De Maddalena), 208 bottom (Michael Siluk), 211 top (Jack Pokoj), 214 (Love Lego), 215 (slowmotiongli), 216/217 (Tory Kallman), 219 top (NFKenyon), 220 top (ChameleonsEye), 221 (Serena Kelly)